A Neuro-Psychoanalytical Dialogue for Bridging Freud and the Neurosciences

Sigrid Weigel • Gerhard Scharbert
Editors

A Neuro-Psychoanalytical Dialogue for Bridging Freud and the Neurosciences

Editors
Sigrid Weigel
Zentrum für Literatur- und Kulturforschung
 (ZfL)
Berlin, Germany

Gerhard Scharbert
Institute for Cultural History and Theory
Humboldt University of Berlin
Berlin, Germany

Assistant Editor
Christine Kutschbach
Zentrum für Literatur- und Kulturforschung
 (ZfL)
Berlin, Germany

The conference preceding this publication, and its proceedings presented here were supported by the German Federal Ministry of Research and Education (BMBF), Reference Nrs. 01UG0712 and 01UG1412.

ISBN 978-3-319-17604-8   ISBN 978-3-319-17605-5 (eBook)
DOI 10.1007/978-3-319-17605-5

Library of Congress Control Number: 2015940404

Springer Cham Heidelberg New York Dordrecht London
© Springer International Publishing Switzerland 2016
This work is subject to copyright. All rights are reserved by the Publisher, whether the whole or part of the material is concerned, specifically the rights of translation, reprinting, reuse of illustrations, recitation, broadcasting, reproduction on microfilms or in any other physical way, and transmission or information storage and retrieval, electronic adaptation, computer software, or by similar or dissimilar methodology now known or hereafter developed.
The use of general descriptive names, registered names, trademarks, service marks, etc. in this publication does not imply, even in the absence of a specific statement, that such names are exempt from the relevant protective laws and regulations and therefore free for general use.
The publisher, the authors and the editors are safe to assume that the advice and information in this book are believed to be true and accurate at the date of publication. Neither the publisher nor the authors or the editors give a warranty, express or implied, with respect to the material contained herein or for any errors or omissions that may have been made.

Cover credits:
Cloud Photo © Sigrid Weigel.
Iceberg Photo © Cornelius20/Dreamstime.com.

Printed on acid-free paper

Springer International Publishing AG Switzerland is part of Springer Science+Business Media (www.springer.com)

Preface

The chapters of this book are partially comprised of the ongoing conversations among an interdisciplinary group of scholars and researchers who collaboratively investigate the changing, current perspectives on neuropsychoanalysis as well as the historical and epistemological preconditions behind it. They are based on an exchange between three different bodies of knowledge: *psychoanalysis, neuroscience*, and *cultural science* (not to be mistaken for cultural studies).

The volume is further comprised of papers from three symposiums, the first of which, "Freud and Neurosciences. Investigating the Dialogue between Psychoanalysis and Neurosciences," was organized by the editors as well as Christine Kirchhoff (Berlin). It took place at the *Center for Literary and Cultural Research* / Zentrum für Literatur- und Kulturforschung (ZfL) in Berlin, 2010. This symposium marked the beginning of a long-term collaboration supported by the ZfL, members of the *International Neuropsychoanalysis Society* (NPSA), and the *Sigmund-Freud-Institut*, Frankfurt/M. It was followed by two further symposiums, "Empathy. A Neurobiological Capacity and Its Cultural and Conceptual History" (Berlin 2013) and "'As-if'—Figures of Imagination, Simulation and Transposition in the Relation to the Self, Others and the Arts" (Villa Vigoni 2014). The authors of this volume come from diverse backgrounds, including the fields of neuroscience, neuropsychology, psychiatry, psychoanalysis, and cultural science. Several of them are psychoanalysts with training in neuroscience.

A Neuro-Psychoanalytical Dialogue for Bridging Freud and the Neurosciences could not have been finished without the editorial help of Japhet Johnstone and Christine Kutschbach (both ZfL, Berlin). To them, and everybody who supported this project, I am very grateful.

Berlin, Germany   Sigrid Weigel

Contents

1	**Introduction** Sigrid Weigel	1

Part I   The Venture of Neuropsychoanalysis

2	**What Is Neuropsychoanalysis?** Mark Solms and Oliver H. Turnbull	13

Part II   Embodiment as Bridge Between Psychoanalysis and Neuroscience

3	**Enactments in Transference: Embodiment, Trauma and Depression. What Have Psychoanalysis and the Neurosciences to Offer to Each Other** Marianne Leuzinger-Bohleber	33
4	**Embodiment in Simulation Theory and Cultural Science, with Remarks on the Coding-Problem of Neuroscience** Sigrid Weigel	47

Part III   The Unconscious Before Freud and After

5	**Signs and Souls: The Prehistory of Psychoanalytical Treatment in Nineteenth-Century French Psychiatry** Gerhard Scharbert	75
6	**Dreams, Unconscious Fantasies and Epigenetics** Tamara Fischmann	91

Part IV   ReVisions of the Drive in Freud and Neuroscience

7   Beyond the Death Drive: Freud's Engagement with Cell Biology
    and the Reconceptualization of His Drive Theory 109
    Sigrid Weigel

8   Drive and Love: Revisiting Freud's Drive Theory 127
    Yoram Yovell

9   The Island of Drive: Representations, Somatic States
    and the Origin of Drive... 137
    Pierre J. Magistretti and François Ansermet

Part V   Concerns of Psychoanalytical Theory

10 Couch Potato: Some Remarks Concerning
    the Body of Psychoanalysis 151
    Ulrike Kadi

11 "The Medulla Oblongata Is a Very Serious
    and Lovely Object." A Comparison of Neuroscientific
    and Psychoanalytical Theories.. 163
    Edith Seifert

On the Authors 171

Contributors

François Ansermet Department of Psychiatry, Université de Genève, Genève, Switzerland

Tamara Fischmann Sigmund-Freud-Institut, Frankfurt am Main, Germany

Ulrike Kadi Universitätsklinik für Psychoanalyse und Psychotherapie, Medizinische Universität Wien, Wien, Austria

Marianne Leuzinger-Bohleber Sigmund-Freud-Institut, Frankfurt am Main, Germany

Pierre J. Magistretti Laboratory of Neurogenetics and Cellular Dynamics, École Polytechnique Fédérale de Lausanne, EPFL-SV-BMI, Lausanne, Switzerland

Gerhard Scharbert Institut für Kulturwissenschaft, Philosophische Fakultät III, Humboldt-Universität zu Berlin, Berlin, Germany

Edith Seifert Psychoanalytischer Salon Berlin, Berlin, Germany

Mark Solms Department of Psychology, University of Cape Town, Cape Town, South Africa

Oliver H. Turnbull School of Psychology, Bangor University, Bangor, UK

Sigrid Weigel Zentrum für Literatur- und Kulturforschung Berlin, Berlin, Germany

Yoram Yovell The Institute for the Study of Affective Neuroscience (ISAN), University of Haifa, Haifa, Israel

Abbreviations

G.W. Sigmund Freud (1940–1957). *Gesammelte Werke, chronologisch geordnet* (18 Vols. plus 1 Nachtragsband). Anna Freud, Marie Bonaparte, E. Bibring, W. Hoffer, E. Kris & O. Osakower (Eds.). London: Imago.

S.E. Sigmund Freud (1953–1974). *The Standard Edition of the Complete Psychological Works of Sigmund Freud* (24 Vols.). James Strachey & Anna Freud (Trans. & Eds.). London: Hogarth Press.

Stud. Sigmund Freud (1969–1975). *Studienausgabe* (10 Vols. plus 1 Ergänzungsband). Alexander Mitscherlich, Angela Richards & James Strachey (Eds.). Frankfurt am Main: Fischer.

Chapter 1
Introduction

Sigrid Weigel

Abstract The introduction concerns itself with the current constellation of the dialogue between psychoanalysis and the neurosciences, identifies the areas where the most productive exchanges are occurring, and views the ongoing use of the hyphen in the word "neuro-psychoanalysis" as a telling symptom that invites questions about the problems inherent in the dialogues as well as their epistemological premises. This affects not only the various methods and techniques involved in the analysis of the psychological apparatus and its relation to physiological functions, but also, crucially, the terminology employed. The translation of Freud's works plays a central role here, as the transfer of his concepts into English coincides with the empirical methods used in the neurosciences, while the reception of psychoanalysis as a theory on how meaning is constructed concentrates on Freud's purposeful language. From a history of science perspective, the constellations around 2000 and 1900 will thereby be interpreted as a reverse mirror situation: a foundation of psychoanalysis based on neurology, while renouncing the localization-based paradigm (around 1900) and the rediscovery of Freud and the subconscious through neuropsychology, experimental psychology, and neuroscience (around 2000).

Keywords Psychoanalysis • Neurosciences • Neuro-Psychoanalysis • Freud • History of Science

The 'emotional turn' in brain research and the rediscovery of Sigmund Freud, the unconscious, and the relevance of dreams by numerous scholars from experimental psychology, neuropsychology, and neuroscience that have taken place during the last few decades have created a fascinating albeit complicated situation. After "a century of misunderstanding" (Whittle 1999) and "alienation" (Watt 2000) several initiatives (some enthusiastic, some more cautious) formulated perspectives to bridge the long existing gap between psychoanalysis and neurobiological disciplines.

S. Weigel (✉)
Zentrum für Literatur- und Kulturforschung Berlin,
Schützenstr. 18, D-10117 Berlin, Germany
e-mail: direktion@zfl-berlin.org

© Springer International Publishing Switzerland 2016
S. Weigel, G. Scharbert (eds.), *A Neuro-Psychoanalytical Dialogue for Bridging Freud and the Neurosciences*, DOI 10.1007/978-3-319-17605-5_1

More and more projects are emerging with the aim of developing interdisciplinary research designs to undertake studies on basic affects (such as anxiety), long-term investigations of specific psychic phenomena like trauma (Leuzinger-Bohleber et al. 2007), depression, and schizophrenia, and studies on special affective disorders. However, this constellation is not as complementary or reciprocal as it looks at the first glance.

Whereas in the past the attitude of distancing originated primarily from the camp of neuroscience whose scholars criticized psychoanalytical theory for not being scientifically grounded, today skepticism comes more from the psychoanalysis side. After having been neglected for such a long time, many psychoanalysts do not trust the neurosciences' current increasing engagement with questions and subjects that, for nearly a century, belonged exclusively to the purview of analysts. Others perceive the turn to affects by a field such as brain research as a hostile embrace, as an attempt to conquer yet another area of research by a field that had been guided by a strict cognitive approach without any thought to affects for decades.

1.1 Bridges and the Hyphen of Neuro-Psychoanalysis

This volume is not primarily occupied with debating the pro and cons of a convergence of the fields. Instead the interdisciplinary discussion of current perspectives on and paths to bridging psychoanalysis and neuroscience is accompanied by historical examinations into the neurological and biological preconditions of Freud's theory in order to shed some light on the historico-scientific background. The epistemological situation of current neuro-psychoanalytical approaches is structured by a cross-over of the two parties involved, given that the situations around 1900 and around 2000 form a kind of reverse mirror constellation. As is well known, Freud generated psychoanalysis as a new approach by turning away from the anatomical foundations of neurology, after having practiced as a successful neurologist for about two decades. His new approach thus emerged as a literal off-spring of neurology. Against the background of this founding scene, the historico-scientific constellation at stake reads as follows: Whereas Freud reached the limits of empirical methods, failed in his ambitious attempt of establishing a biological theory of memory based on neurological explanations (as elaborated *in Project for a Scientific Psychology/Entwurf einer Psychologie,* 1895), and finally turned to psychoanalysis, today many scholars of neuroscience with advanced technology at their disposal for the study of neurological and biochemical brain functions have reached the limits of a mere cognitive approach and are rediscovering questions developed and defined by psychoanalysis. This constellation brings with it central epistemological problems as well as numerous misunderstandings that are inscribed in the process of finding a fruitful conflation of the different bodies of knowledge.

For many scholars versed in metapsychology, it is difficult or even impossible to recognize the Freudian questions, ideas, and key concepts when they encounter them in the guise of recent neurobiological terminology. Their disconcertion is in part caused by the reconceptualization of the subject matter within a totally different episteme and a terminology based on anatomical mappings of brain areas (and its lesions), of somatic markers, neurological functions, the role of neurotransmitters. The brain maps are, for example, sometimes accompanied by evolutionary interpretations on the behavior patterns of newborns and infants. But the impression of being confronted with unfamiliar concepts is also due to the fact that the majority of psychoanalytical theory from the last century casted a shadow over the biological groundings of Freud's theory. One symptom of the prevalence of psychoanalysis and its discontent with 'biology' is the controversial positions regarding Freud's concept of 'drive' in this field.

The absence of his early neurological writings in Sigmund Freud's *Gesammelte Werke* (edited 1940–1952, complemented 1987) marks a lack in the dominant image of Freud that further coincides with a widespread indifference towards or even negation of the neurological training the founder of psychoanalysis previously underwent. A serious analysis of the epistemological problems inherent in the gap between the fields demands a historico-scientific inquiry of its genesis and development. This includes the question of how far and in which way Freud's metapsychology is influenced or informed by his experience and knowledge as a neurologist. This kind of investigation has started only recently (Guttmann and Scholz-Strasser 1998; first chapter in Kaplan-Solms and Solms 2000). The task of reflecting, elucidating, and overcoming the epistemological reasons for a century of silence between the two fields is a difficult enterprise that will take a long time and a great deal of devotion, mutual endurance, and scientific curiosity. This volume is understood as a small contribution to a great undertaking. One of its propositions is to incorporate the work of scholars from the humanities engaged with historico-scientific perspectives on the conceptualization of the 'psychical apparatus' and the crucial epistemological questions involved therein.

The project of bridging psychoanalysis and neuroscience demands more than simply developing a dialogue and exchanging mutual praise; it needs to address the question, "What we can learn from a century of misunderstanding?" as Paul Whittle from experimental psychology properly formulated in one of the first issues of the journal *Neuro-Psychoanalysis*, founded in 1999 by Mark Solms, Edward Nersessian, and colleagues (Whittle 1999). To analyze the historico-scientific preconditions of this enterprise requires, on the one hand, investigating and uncovering the fate of neurology and the role of biological elements in Freudian psychoanalysis and, on the other, to examining the vicissitudes of psychoanalytical concepts and questions when they enter today's neuroscience laboratories. The success of the fascinating and promising prospect to bridge the fields depends on the "awareness that psychoanalysis is completely based on intersubjectivity whereas the neurosciences are based on the relationship between the subject and his or her object of interest" (Mancia 2006, 2).

Contemporary research conditions and tools used to study the development, functions, and expressions of the psychical apparatus are not just determined by a historical distance to Sigmund Freud's time, nor should these changes be characterized as mere scientific progress due to advanced technology and tools "to look into the brain," as if Freud's dream of 'one day having other possibilities at hand' has now been fulfilled. This conclusion is tempting, since Freud repeatedly reflected on an altered research setting in future, as for example in *Beyond the Pleasure Principle* (1920): "The deficiencies in our description would probably vanish if we were already in a position to replace the psychological terms by physiological or chemical ones" (S.E. XVIII, 60). However, the history of science in the twentieth century has excluded and discriminated against crucial concepts of Freudian theory in part based on the development of the very physiological and chemical concepts that Freud alludes to. Science is no longer the same; nowadays it finds itself in the wonderful and comfortable position of having those means at hand that Freud could only dream about. Therefore, the question has to be addressed as to what extent those exclusions are inscribed as pivotal aspects of the neuroscientific approach. The rediscovery of the *unconscious* by neuroscience for example cannot just mean reconceptualizing Freudian concepts by means of new research technology and translating them into the terminology of neurobiology. It needs to include a critical reflection of the fundamental tenets of neuroscience.

In this respect the discovery of the indispensable role of emotions for brain activity (Damasio 1994; see also *The Emotional Brain*, Le Doux 1996) has prepared conditions of possibility for bridging the gap to psychoanalysis. Equally important was the discovery of somatosensory *mirror neurons* that provide the neurobiological requisite for a resonance mechanism between individuals (Rizzolatti and Sinigaglia 2008) and, additionally, the ensuing formulation of the theory of *embodied simulation* (Gallese 2003, 2009). These steps in brain research prepared the ground for a further and deeper convergence because it triggered a shift in the whole episteme, especially concerning the image of the human subject. The previous epoch of neuroscience was dominated by a computer model of the brain, which presents the brain as a person's steering center for all bodily and cognitive functions. Recently, the paradigm of mirror neurons introduced concepts of intersubjectivity into the field, thereby opening up the idea of a shared space of affects and perceptions. In the wake of this caesura, the neurobiological model of the human subject and its psychical apparatus approaches more and more the basic tenets of psychoanalysis whose theory and clinical practices have been based in intersubjectivity from the very beginning.

The neuroscientific approach of psychoanalysis has invented a *new biology of mind* and raised the demand to verify by experimental method the functional modifications and plasticity of an individual brain caused by therapy (Kandel 2005). In addition, innumerable findings from recent neuroscientific research have confirmed crucial claims from Freud's hypotheses, in particular his ideas on the eminent role of the unconscious and dreams, his emphasis on the early development of the infant and its relationship to the mother's body, his model of how the memory is organized, and his observations on certain psychic disorders (for a survey, see Mancia 2006). Still, the hyphen between the two bodies of knowledge exists although the neuro-psychoanalytical endeavors have reached a stage where collaboration is possible both on the clinical level and in a theoretical engagement with central aspects of the psychical apparatus like dream, memory, etc. The hyphen furthermore indicates unsolved controversies and open questions.

In this respect, a different terminology might highlight irreconcilable differences in the epistemologies that are to some extent still incompatible. Thus, it makes a difference whether one deals with a dual memory system, consisting of *implicit memory* and *explicit memory* (Schacter 1996), or whether one refers to Freud's idea of a dynamic incompatibility and the un-simultaneity of two systems, *perception-consciousness* (pcs.) and *the unconscious* (ucs.). And it makes a difference if one speaks of a psychical apparatus or a mental apparatus.

1.2 Terminology Counts

Two phrases are currently at the center of the prevalent terminology being used to integrate Freud's ideas seemingly unimpeded into the contemporary scientific language: the concept of a *mental apparatus* and the widely used formula of 'mind and brain'. Actually, the terms *mental* and *mind* signify a nearly imperceptible shift in the meaning of key concepts and questions stemming from Freud's conceptualization of psychoanalysis: a shift from the realm of the psyche to that of the mind. If words like *Seele* (soul), *Psyche* (psyche), and *psychischer Apparat* (psychic apparatus) get translated as the 'mental apparatus,' they subsequently get transformed into a conceptual field that no long differentiates appropriately between cognitive and psychic aspects. The shift in question remains implicit and nearly imperceptible because the language in the current discourse on Freud and neuroscience is but the last step in a long line of continuous modifications of meaning in the process of translation.

When dealing with Freud within a neurobiological context today, the terminology at hand is the result of multiple translations over the years: translations between different languages, different disciplines or bodies of knowledge, and between different historical states of language. The displacement of meaning in Freudian key concepts reaches back to the translated editions of his writings. The often debated difficulties of translating Freud into English, French, or any other language have motivated several initiatives of new translations and re-translations. Many scholars, analysts, and translators in different countries are engaged in this challenging venture; the most important translation project out to date is the revision of the *Standard Edition* accompanied by a four-volume edition of Freud's *Complete Neuroscientific Papers* by Mark Solms that will finally complete Freud's oeuvre. This complex issue shall not be rehashed here in the detail; what is worth considering in our context is the transfer of Freud's *seelischer* or *psychischer Apparat* to *mental apparatus*, his *Seele*, *Psyche*, or *Seelenvermögen* to *mind*, his *Trieb* to *instinct* and the like; the metamorphosis of these terms directed the path of psychoanalytical ideas towards biomedical and scientific terminologies and, in this way, supported their adoption by neuroscience. Since the present working language of scholars in (natural) sciences is English, the linguistic transfer enabled the approximation of neuropsychology to psychoanalysis. But this advantage is, at the same time, based on a certain neutralization of Freudian terms; it profits from the withdrawal of a signifying surplus to a strict scientific terminology characteristic of Freud's use of language; this surplus is lost in translation. Whereas the mental and the mind can be assimilated to the episteme of experimental research, the central term for the Freudian subject, *psyche*, still owns something of its fluttering origin: the Greek *psyché* being the word for both soul and butterfly.

When Freud decided to concentrate on psychic disorders that are not caused by or based in anatomical lesions of the nervous system, when he decided to study psychic disorders, neuroses, and hysteria, he not only distanced himself from the neurological localization paradigm (Solms and Saling 1986) of his time, but he also substituted the repertoire of clinical *syndromes* by *symptoms*. Whereas the former is defined as an identifiable image of a certain disease, Freud conceptualizes the latter as a product of a psychical processing or 'working out', as the site where the drive realizes its final expressive aim, as Starobinski puts it: "als Realisationsort der expressiven Endzwecke des Triebs" (Starobinski 1987, 23). Freud calls the (hysteric) symptom a *mnemic symbol* (*Erinnerungssymbol*)—and a symbol always stands for something else. This 'standing for' is a basic characteristic of the language of the unconscious and as such has provided at least one of the main aspects of psychoanalysis responsible for the strong engagement of scholars from the humanities with Freud's theory, both as a general theory for dealing with anthropology, cultural phenomena, and historical experience and as a repertoire of specific figures of cultural interpretation like hysteria, melancholy, trauma, etc. (see, for example, Bronfen et al. 1999 on trauma). In terms of epistemology this 'language' is positioned at the threshold between physiological phenomena and psychical meaning, between the quantitative paradigm and qualities, between data and the affective state of the individual: it is based in physiological modifications of the nervous system, but full of meaning that goes far beyond.

On the one hand, it is promising and fascinating to see how today's brain technology is able to visualize certain Freudian ideas, as, for example, the plasticity formed by 'synaptic traces' (Ansermet and Magistretti 2004) does in relation to Freud's concept of facilitation (*Bahnung*), which he described as neuronal paths in the psychic apparatus that form the memory, namely, the memory represented by the facilitations existing between the ψ-neurons ("Das Gedächtnis ist dargestellt durch die zwischen den ψ-Neuronen vorhandenen Bahnungen." *Entwurf einer Psychologie*, 1895, G.W. Ergänzungsbd., 308). On the other hand, one has to reflect upon the fate of the surplus of connotations belonging to an idea like *Erregungsvermögen* in comparison to "physiological modification of the nervous system" (*Hysteria*, 1888, S.E. I, 41).

Georges-Arthur Goldschmidt in *Quand Freud voit la mer* (1988) argues that Freudian psychoanalysis is based in a specific use of the German language. However, this argument does not refer to the national character of the language but rather to certain aspects or possibilities of it such as its strong *sensual quality* and the commonness of *topic figures*[1] that Freud benefited from in the formation of his pictorial ideas, frequently mistaken as metaphors. But Freud's linguistic images are not metaphors in the strict sense of having figurative meaning because there are no concepts at hand for which they are meant to stand. Against the background of the theory of metaphorology, they could best be described as 'absolute metaphors', images in language that cannot be dissolved into terminology ("nicht in Begrifflichkeit aufgelöst werden können," Blumenberg 1960, 11) but rather are used as figurative concepts. To mention just two passages where Freud reflects on his *Bildsprache*: In his *Introductory Lectures on Psycho-Analysis/Vorlesungen zur Einführung in die Psychoanalyse* (1916), he states that he can provide the audience only a pictorial depiction (*bildliche Schilderung*) instead of a properly ordered topographical-dynamic description (*topisch-dynamisch geordnete Beschreibung*) when presenting melancholia as a position from which it is possible to receive insights into the inner structure of affection[2] (Stud. I, 412 f.). And in *Civilization and Its Discontent/Das Unbehagen in der Kultur* (1930), Freud admits that he is momentarily not able to characterize the kind of satisfaction artists and scholars have in their work (due to the displacement of libido and the drive's aims) in a metapsychological way but can only say "bildweise", which means using words like images: their satisfaction is "finer and higher" (*feiner und höher*, Stud. IX, 211). And even when Freud uses diagrams in order to offer a scientific representation of his theses, these diagrams are far from standard illustrations from physiognomy or biology. An analysis of the few images Freud includes in his analytical publications[3] shows that Freud mainly uses abstract schematic diagrams which do not encourage drawing associations to physiognomic morphology or anatomy. This means that even in diagrammatical representations Freud used rhetoric visualizations "because the psychical apparatus can only be described by language and not represented by pictures" (Borck 1998, 84–85).

The position of Freud's concepts on the threshold between physiognomy and affective meaning, or rather the symbolic, provides a wonderful bridge between neurobiology and the realm of interpretation. But the abyss beneath the bridge is full of complicated epistemological problems, first and foremost the unsolved question of how to translate between empirical approaches and a body of knowledge based on interpretation and understanding.

---

[1] Goldschmidt refers especially to the register of prefixes (like *vor-, un-, ent-, wieder-*) and their concrete spatial or temporal sense.
[2] This statement concerns the melancholiac's "narcissistic identification" with the lost object that results in blaming oneself instead of addressing reproaches against the latter.
[3] Cornelius Borck counts eight in comparison to the twenty to be found in Freud's neuroanatomical works (Borck 1998).

1.3 Present Sites of Bridging and Controversy

In addition to general reflections on neuro-psychoanalytical perspectives, its chances, and difficulties, this volume presents some exemplary concepts that are currently at the center of a neurobiological reformulation of, or neuroscientific convergence with, psychoanalysis: embodiment, the unconscious, and drive.

Drive belongs to the most complicated and controversial concepts of Freudian psychoanalysis. The fact that Sigmund Freud describes the drive as a liminal concept (*Grenzbegriff*) between the soul and the somatic in *Instincts and their Vicissitudes/Triebe und Triebschicksale* (1915) can be regarded as the reason for the problems many psychoanalysts as well as scholars from the humanities occupied with Freud's theory have with this concept. Skepticism against the 'drive' is a symptom of the vicissitude of the concept itself within a century of psychoanalytical theory. Its description by Freud reaches quite far into the realm of the biological and neurological foundations of the psychic apparatus theory. Freud himself defined drive as a representative of bodily stimuli within the soul. Therefore, the drive brings with it physiological functions into the psychic apparatus and, with respect to research, it is the biological representative within psychoanalysis. When Freud states that the drive determines the amount of required work that is imposed to the soul because of its relation to corporeality ("das Maß der Arbeitsanforderung, die dem Seelischen infolge seines Zusammenhanges mit dem Körperlichen auferlegt ist" (Stud. III, 85)), this means that the idea of drive consequently determines the amount of engagement with physiology or biology that is demanded by psychoanalysis.

The *unconscious*, in contrast, is an idea that is positioned at the farthest most conceivable distance from anatomy, biology, and neurobiology. This is the reason why *dreams*, as the most prominent phenomenon in Freud's register of the language of unconscious, had been banned for decades from the agenda of brain research and neuroscience. But recently neuroscientific research was able to confirm single aspects of Freud's hypotheses on the relationship between sleep and dreaming, as for example the idea that dreaming functions as a protector of sleep, as "Wächter des Schlafs" (*Interpretation of Dreams/Traumdeutung*, 1900, G.W. II, 239; Solms 1995b) and his conviction that the inhibitory activity (i.e., censorship) is lowered during dream sleep (Yu 2000). However, it was not until after finding out that different brain areas are 'responsible' for REM-sleep and dreaming (Solms 2000) that dreams have been rehabilitated as serious subjects of scientific research. After this step scholars of neuroscience started their search for ways and methods to explore how to empirically study dreams. Understanding the two sides of dreams (the dynamic, irregular neurologic activity as the physiological side of dreaming and the ephemeral images of dreams as the other side) may be regarded as one of the most difficult test cases of neuro-psychoanalysis.

Conversely, neuroscience's turn to concepts of *embodied simulation* (Gallese 2009) deposes the brain from its role as a control and steering center of all movements, functions, perceptions, and articulations of the human body and, in this way, provides access to other concepts of expression and memory beyond the cognitive mind. These concepts have long been discussed under the heading of *embodiment* (Fuchs et al. 2010; Leuzinger-Bohleber et al. 2013). In this way neuroscience combines psychoanalysis's body of knowledge as based in experience and observation with the conversion of psychic agitation into bodily symptoms as well as with other embodied forms of the language of the unconscious in memory. In addition, the increasing acknowledgement of the bodily and mimetic modes of intersubjectivity prepares the ground for a dialogue with the humanities, in which a broad and differentiated register of affects (expressed through gestures, body language, and the like) have been objects of study for years.

This volume is framed by a strong statement in favor of neuro-psychoanalysis by Mark Solms and Oliver Turnbull, in which the authors describe the developments, opportunities, and achievements of the new field. This *pro*-prologue has its *con*-parallel in an epilogue of skeptical reflections from a psychoanalytic perspective by Ulrike Kadi and Edith Seifert, who argue for the autonomy of key concepts from psychoanalysis. In between, thematic sections structure the book, beginning with *embodiment* as one of the most promising bridges between the two fields in question, from a psychoanalytic angle by Marianne Leuzinger-Bohleber and from a cultural science perspective by Sigrid Weigel. Then, the section on the *unconscious and dreams* consists of a historical investigation in nineteenth-century psychiatry before Freud's debut by Gerhard Scharbert and a report of an interdisciplinary research project between neuroscience and psychoanalysis on dreams accompanied by theoretical arguments by Tamara Fischmann. In the section on the concept of *drive*, an analysis of the biological references in Freud's concept by Sigrid Weigel is followed by a neurobiological revision of the concept with emphasis on infants' relations to their mothers by Yoram Yovell and a reconceptualization of the drive concept by the interdisciplinary team François Ansermet and Pierre Magistretti.

References

Ansermet, F., & Magistretti, P. (2004). *À chacun son cerveau: Plasticité neuronale et inconscient*. Paris: Odile Jacob.
Blumenberg, H. (1960). *Paradigmen zu einer Metaphorologie*. Bonn: Bouvier.
Borck, C. (1998). Visualizing nerve cells and psychical mechanisms: The rhetoric of Freud's illustrations. In G. Guttman & I. Scholz-Strasser (Eds.), *Freud and the neurosciences. From brain research to the unconscious* (pp. 57–86). Vienna: Verlag der Österreichischen Akademie der Wissenschaften.
Bronfen, E., Erdle, B. R., & Weigel, S. (1999). *Trauma: Zwischen Psychoanalyse und kulturellem Deutungsmuster*. Köln: Böhlau.
Damasio, A. (1994). *Descartes' error: Emotion, reason, and the human brain*. New York: Putnam.
Damasio, A. (1999). *The feeling of what happens: Body and emotion in the making of consciousness*. New York: Harcourt.
Freud, S. (1888). *Hysteria*. S.E. I. 37–59.
Freud, S. (1895). *Entwurf einer Psychologie*. G.W. Nachtragsband, 373–486.
Freud, S. (1900). *Traumdeutung*. G.W. II/III, 1–642.

Freud, S. (1915). *Triebe und Triebschicksale*. G.W. X, 209–232.
Freud, S. (1916). *Vorlesungen zur Einführung in die Psychoanalyse*. Stud. I, 33–445.
Freud, S. (1920). *Beyond the pleasure principle*. S.E. XVIII, 1–64.
Freud, S. (1930). *Das Unbehagen in der Kultur*. Stud. IX, 191–270.
Fuchs, T., Sattel, H., & Henningsen, P. (Eds.). (2010). *The embodied self: Dimensions, coherence and disorders*. Stuttgart: Schattauer.
Gallese, V. (2003). The roots of empathy: The shared manifold hypothesis and the neural basis of intersubjectivity. *Psychopathology, 36*, 171–180.
Gallese, V. (2009). Mirror neurons, embodied simulation, and the neural basis of social identification. *Psychoanalytic Dialogues, 19*(5), 519–536.
Goldschmidt, G.-A. (1988). *Quand Freud voit la mer*. Paris: Buchet-Chastel.
Guttmann, G., & Scholz-Strasser, I. (Eds.). (1998). *Freud and the neurosciences: From brain research to the unconscious*. Vienna: Verlag der Österreichischen Akademie der Wissenschaften.
Kandel, E. R. (2005). *Psychiatry, psychoanalysis and the new biology of mind*. Washington, DC: American Psychiatric Pub.
Kaplan-Solms, K., & Solms, M. (2000). *Clinical studies in neuro-psychoanalysis: Introduction to a depth neuropsychology*. Madison: International Universities Press.
Le Doux, J. (1996). *The emotional brain: The mysterious underpinnings of emotional life*. New York: Simon and Schuster.
Leuzinger-Bohleber, M., Roth, G., & Buchheim, A. (2007). *Psychoanalyse – Neurobiologie – Trauma*. Stuttgart/New York: Schattauer.
Leuzinger-Bohleber, M., Emde, R. M., & Pfeifer, R. (Eds.). (2013). *Embodiment: Ein innovatives Konzept für Entwicklungsforschung und Psychoanalyse*. Göttingen: Vandenhoeck & Ruprecht.
Mancia, M. (Ed.). (2006). *Psychoanalysis and neuroscience*. Berlin: Springer.
Rizzolatti, G. & Sinigaglia, C. (2008). *Mirrors in the brain. How our minds share actions and emotions*. (F. Anderson, Trans.). Oxford: Oxford University Press (ital. original 2006: *So quel que fai, il cervello che agisce e i neuroni specchio*).
Schacter, D. L. (1996). *Searching for memory: The brain, the mind, and the past*. New York: Basic Books.
Solms, M. (1995a). Is the brain more real than the mind? *Psychoanalytic Psychotherapy, 9*, 107–120.
Solms, M. (1995b). New findings on the neurological organisation of dreaming: Implications for psychoanalysis. *Psychoanalytic Quarterly, 64*, 43–67.
Solms, M. (1996a). Towards an anatomy of the unconscious. *Journal of Clinical Psychoanalysis, 5*, 331–367.
Solms, M. (1996b). Was sind Affekte? *Psyche, 50*, 485–522.
Solms, M. (2000). Dreaming and REM sleep are controlled by different brain mechanisms. *Behavioral and Brain Sciences, 23*, 843–850.
Solms, M., & Saling, M. (1986). On psychoanalysis and neuroscience: Freud's attitude to the localizationist tradition. *The International Journal of Psycho-Analysis, 67*(4), 397–416.
Solms, M., & Turnbull, O. (2002). *The brain and the inner world: An introduction to the neuroscience of subjective experience*. London: Karnac.
Starobinski, J. (1987). *Kleine Geschichte des Körpergefühls*. Konstanz: Universitätsverlag.
Starobinski, J., Grubich-Simitis, I., & Solms, M. (2006). *Hundert Jahre Traumdeutung von Sigmund Freud: Drei Essays*. Frankfurt: Fischer.
Watt, D. (2000). The dialogue between psychoanalysis and neuroscience: Alienation and reparation. *Neuropsychoanalysis. A Journal for Psychoanalysis and the Neurosciences, 2*(2), 183–192.
Whittle, P. (1999). Experimental psychology and psychoanalysis: What we can learn from a century of misunderstanding. *Neuropsychoanalysis. A Journal for Psychoanalysis and the Neurosciences, 1*(2), 233–245.
Yu, C. K.-c. (2000). Clearing the ground: Misunderstanding of Freudian dream theory. *Neuropsychoanalysis. A Journal for Psychoanalysis and the Neurosciences, 2*(2), 212–213.

Part I
The Venture of Neuropsychoanalysis

Chapter 2
What Is Neuropsychoanalysis?

Mark Solms and Oliver H. Turnbull

Abstract The authors examine the historical, philosophical, and scientific foundations of what they term the "inter-discipline" of neuropsychoanalysis. With the support of historical evidence, they unravel how the traditional strict separation of psychoanalysis and the neurosciences, as still practiced by certain scientists in both fields, is grounded in a misreading of Freud: Rather than advocating such a separation *on principle*, Freud developed his purely psychoanalytical method for *pragmatic* reasons—neuroscience in his time simply was not advanced enough to yield fruitful results. While psychoanalysis continued to use subjectivity to explore the internal perception of the mental apparatus, neuroscience developed tools to study the physical realization of the mental apparatus in the brain. The authors argue that a position of correlation between the two modes is likely to yield stronger results than a single-track focus. Far from constituting a new school of psychoanalysis, neuropsychoanalysis provides a link that integrates research being conducted along the psychoanalysis/neuroscience boundary.

Keywords History of neuropsychoanalysis • Clinico-anatomical method • Neuroscientific technology • Dual-aspect monism • Metapsychology

The first formal use of the term 'neuropsychoanalysis' was in 1999, when it was introduced as the title of the journal *Neuropsychoanalysis*. Plainly the relationship between psychoanalysis and neuroscience is much older than the term. In the dozen years since the word 'neuropsychoanalysis' was first used, it has been employed in a number of different ways, for different purposes, by different people. This chapter serves to briefly survey some of this complexity and in the process to sketch out the intended scope of the field.

Based on a paper presented at the 10th International Neuropsychoanalysis Congress, Paris. A version of this article was published in *Neuropsychoanalysis 13*, 133–45 (2011).

M. Solms (✉)
Department of Psychology, University of Cape Town, Cape Town, South Africa
e-mail: mark.solms@uct.ac.za

O.H. Turnbull
School of Psychology, Bangor University,
Adeilad Brigantia, Penrallt Road, Gwynedd, LL57 2AS Bangor, UK
e-mail: o.turnbull@bangor.ac.uk

There are two major limitations to this account. The first is that we can speak only for ourselves, and thus describe what *we* think 'neuropsychoanalysis' is—and ought to be. Nevertheless, we may claim a certain privilege in this respect by virtue of us having invented the term. Secondly, we aim to speak only of the absolute basics of the discipline, to address only the foundational issues. There is much that is both pertinent and interesting that we will not even attempt to address here.

We will address the question "what is neuropsychoanalysis?" under four headings: (1) the historical foundations of neuropsychoanalysis, (2) its philosophical foundations, (3) its scientific foundations, and lastly (4) we will discuss what neuropsychoanalysis is *not*.

2.1 Historical Foundations of Neuropsychoanalysis

When we speak of the historical foundations of neuropsychoanalysis, we must, of course, begin with Freud. In doing so we are also addressing the question as to whether or not *neuro*psychoanalysis is really a legitimate part of psychoanalysis. The alternative view is that it is somehow a foreign body in our midst, perhaps even something fundamentally *anti*-psychoanalytical.

In relation to this question, Freud's attitude to the issue is of paramount importance. If neuropsychoanalysis is legitimately part of what Freud conceived psychoanalysis to be, it places the inter-discipline of neuropsychoanalysis in a strong position with respect to its parent discipline. It was Freud, after all, who invented psychoanalysis. Happily, his view on the matter was very clear and also consistent throughout his life. He was, of course, a neuroscientist and a neurologist for the first two decades of his professional life (Solms 2002; Solms and Saling 1986). Throughout this later psychoanalytic work he had a specific scientific programme in mind, largely continuous with his earlier neuroscientific work, albeit shaped by the limitations of the scientific methods and techniques available to him at that time (for more on this topic see Solms and Saling 1986; Solms 1998; Turnbull 2001).

Freud's programme was to map the structure and functions of the human mind, and naturally he recognised that these were intimately related to the structure and functions of the human brain. However, as regards these relationships, he consistently argued that the brain sciences of his time did not have the *tools* (in both conceptual and technical terms) necessary for exploring them. He therefore shifted to a purely psychological method—a shift that he reluctantly saw as a necessary expedience. Just a few quotations illustrate this position. In his essay *On Narcissism/Zur Einführung des Narzißmus* (1914) he discussed the eventual possibility to discover the "organic substructure" of psychological functions:

> [W]e must recollect that all our provisional ideas in psychology will presumably some day be based on an organic substructure [...]. We are taking this probability into account in replacing the special chemical substances by special psychical forces. (S.E. XIV, 78)

And six years later in *Beyond the Pleasure Principle/Jenseits des Lustprinzips* (1920) he speculated upon the meaning of the chemical substances. He imagined that a biological answer to his questions would emerge "in a few dozen years". This estimation was not far from the truth:

> The deficiencies in our description would probably vanish if we were already in a position to replace the psychological terms by physiological or chemical ones. [...] Biology is truly a land of unlimited possibilities. We may expect it to give us the most surprising information and we cannot guess what answers it will return in a few dozen years to the questions we have put to it. (S.E. XVIII, 60)

There are many such statements throughout Freud's work. All reveal, firstly, that he viewed the separation of psychoanalysis from neuroscience as a *pragmatic* decision. Secondly, he was always at pains to clarify that progress in neuroscience would have the inevitable result that *at some time in the future* the neurosciences will advance sufficiently to make the gap bridgeable.

What were the methodological limitations that Freud encountered at his time? The main neuroscientific tool then available was the clinico-anatomical method, based on the clinical investigation of patients who had suffered focal brain lesions (Finger 1994); that is to say, studying how different functions of the mind were altered by damage to different parts of the brain. It was effectively the *only* method available for studying mind-brain relationships (though Freud's later years did briefly overlap with early developments in psychopharmacology; see Finger 1994).[1] However, Freud regarded the clinico-anatomical method as unsuitable for his purposes, despite having used it himself in his pre-analytic work. Freud set out his arguments against clinico-anatomical localization in his 1891 book on the aphasias, which demonstrates how sophisticated his mastery of that method was and also its limitations (see Shallice 1988, 245–247, for a modern appreciation of Freud's early neuropsychological investigations).

Within five years Freud rejected neuropsychology altogether (Solms 2001), as he made the transition into psychoanalysis. He did so for several reasons. Firstly, he recognised that the mind is a dynamic thing. It was Freud's emphatic view, even as a neurologist and as author of his critical work *On Aphasia/Zur Auffassung der Aphasien* (1891), that the mind [*Seelenvermögen*] was not made up of static modules or boxes connected by arrows. Instead, he saw the mind as comprising dynamic, fluid processes. Secondly, Freud observed that the mind consisted in far more than consciousness; there was, beneath consciousness, a vast invisible sub-structure, the workings of which had to be explored and understood and exposed before they could ever be 'localized'—before we would ever be able to make sense of the volitional brain. The aim of psychoanalysis then became to develop a method, and ultimately to derive from that method a theory (and a therapy), which would enable science to expose the dynamics of the unconscious mind. It is widely known that Freud then proceeded to use this purely clinical method free from neuroscientific constraints from around 1895 until 1939. This pioneering work left us a great legacy,

---

[1] "The future may teach us to exercise a direct influence, by means of particular chemical substances, on the amounts of energy and their distribution in the mental apparatus" (S.E. XXII, 182).

including a series of theoretical models of the basic organisation of the mind, which we now refer to as 'metapsychology'.

For Freud, this was a necessary first step toward a future neuropsychology of mental *dynamics*:

> The psychical topography that I have developed [...] has nothing to do with the anatomy of the brain, and actually only touches it at one point.[2] What is unsatisfactory in this picture – and I am aware of it as clearly as anyone – is due to our complete ignorance of the *dynamic* nature of mental processes. (S.E. XXIII, 97)

Some psychoanalysts, failing to recognize the neurological underpinnings of Freud's ideas and thus misreading his work, argue that the theoretical work of psychoanalysis must continue to remain aloof from neuroscience forever. They claim we must avoid using neuroscientific methods of any kind, no matter how far these advance, and cling to our exclusively clinical, psychological approach. These are authors who question "whether the study of [neuroscience] contributes in any way to the understanding or development of psychoanalysis as theory or practice" and "whether neuroscience is of value to psychoanalysis *per se*" (Blass and Carmeli 2007, 34). The proponents of this view appear to form a diminishing minority— fortunately so in our opinion (British Psychoanalytical Society 2008). But we must acknowledge that there are still some colleagues who believe that psychoanalysis has nothing to learn from neuroscience *in principle*. (Oddly, however, they do seem to think that neuroscience has something to learn from psychoanalysis!) Independently of this theoretical, or ideological, question, there remains the *technical* question as to whether neuroscience has developed sufficiently as a discipline to allow it to make an adequate contribution to psychoanalytic theory, which is to say whether the methodological limitations (and the related limitations of neuropsychological knowledge that Freud referred to) still remain. Stepping back, it is clear that there have been huge technical and methodological advances in the neurosciences over the last several decades. Here is but the briefest historical summary:

Electroencephalography (EEG) was introduced in the 1930s (Berger 1929), though it was not fully exploited until after the war. This represented the beginning of a capability, initially rather crude, to measure and observe dynamic aspects of brain activity under changing functional conditions. The later development of event-related potentials (ERPs) in the 1960s (Sutton et al. 1965; Sutton et al. 1967; Walter et al. 1964; see Luck 2005) offered substantial advances over the basic EEG technique by virtue of experimental control and averaging procedures. The recent development of magnetoencephalography (MEG) represents a further substantial advance, allowing us to study the neural dynamics associated with mental events at the millisecond level and with increasing anatomical precision.

In another domain, after the Second World War, there were tremendous developments in neuropsychology, involving the lesion method in a new way, which adapted its inherent limitations to the dynamic nature of the mind. Alexander Luria, in particular, developed a method known as 'dynamic localisation' (Luria 1966, 1973;

---

[2] Freud was referring here to the *conscious* part of the mind, to what he called the system Pcpt.-Cs.

see Kaplan-Solms and Solms 2000, 39–4; see also Solms and Turnbull 2002, 64–66). This method permitted the investigator to identify constellations of brain structures that interact to form dynamic functional systems, where each structure contributes an elementary component function to the complex psychological whole. On this basis, modern neuropsychology has a well-developed understanding of most mental functions. This applies especially to cognitive functions.

Further enormous technical advances followed the advent of computerised tomography in the 1970s, which made it possible to identify the precise location of a brain lesion while the patient was still alive. This was followed by Magnetic Resonance Imaging (MRI). And from the 1990s onward, functional neuroimaging (functional Magnetic Resonance Imaging [fMRI], Position Emission Tomography [PET] and Single Photon Emission Computed Tomography [SPECT]) made it possible to *directly observe* neurodynamic processes under changing psychological conditions. It is now also possible to deliver temporary, short-acting 'lesions' to neurologically intact research participants, either through sodium amytal injection (which was first introduced in the 1940s) or through magnetic pulses delivered to the outside of the skull via *Transcranial* Magnetic Stimulation (TMS; which has been readily available since the 1990s). Innumerable other technologies also exist, ranging from stimulation of the cortical surface in neurosurgical operations (Penfield 1952), to deep brain stimulation (DBS; Mayberg et al. 2005), and to psychopharmacological probes (Ostow 1962), to mention only the most obvious examples.

Even this brief summary demonstrates that we *do* now have neuroscientific methods that enable us to study the dynamic nature of the mind and to identify the neural organisation of its unconscious substructure. Each of these methods has its limitation, as all methods do, and there are undoubtedly many future advances to come. But the landscape of scientific enquiry in this domain has, certainly, *radically* changed since Freud's lifetime. For this reason, it seems entirely appropriate to reconsider whether we might now attempt to map the neurological basis of what we have learnt in psychoanalysis about the structure and functions of the mind, using neuroscientific methods available to us today. Freud, in our opinion, would have considered this a welcome and wholly legitimate development of the work that he pioneered.

2.2 Philosophical Foundations of Neuropsychoanalysis

If we are to correlate our psychoanalytical models of the mind with what we know about the structure and functions of the brain, we are immediately confronted with the philosophical problem of how mind and brain relate, i.e., with a difficult aspect of the 'mind-body problem'. This opens huge philosophical questions. Are we *reducing* the mind to the brain, are we *explaining away* the mind, or are we merely correlating mind and brain? And if we are merely correlating them, what is the causal basis of this apparently compulsory correlation? Is the relationship hierarchical, whereby psychoanalysis studies mere epiphenomena of the brain? Or is the mind *an emergent property* of the brain? (See Solms 1997a or Solms and Turnbull 2002, 45–66 for a basic review of these issues.) It is, of course, terribly important in this field to be clear about one's conceptualisation of the relationship between the mind and brain. We favour a conceptualisation (shared by Freud) which we think neuropsychoanalysis as a whole may be based upon. We call this approach 'dual-aspect monism' (see Solms 1997a or Solms and Turnbull 2002, 56–58). In what follows we will explain this concept in relation to Freud's ideas.

In *Some elementary lessons in psychoanalysis* (1938) Freud says, very clearly in many places, that the actual nature of the mind is unconscious (see Solms 1997a for review). He uses the phrase "the mind [...] *in itself*" [*das Psychische an sich*] (S.E. XXIII, 283, transl. mod.), sometimes, as in *The Unconscious* (1915), referring directly to the philosophy of Kant (S.E. XIV, 171). For Kant, our subjective being, the thing we perceive when we look inwards, is not the mind *in itself*; the mind in itself cannot be perceived directly. We can only know the mind via our phenomenal consciousness, which provides an indirect and incomplete *representation* of the mental apparatus and its workings. The actual ontological nature of the mind is something epistemologically unknowable; it necessarily lies behind (and generates) conscious perception. We can, of course, *infer* its nature from our conscious observations, and thereby 'push back' the bounds of consciousness, which is what the psychoanalytical method seeks to do. Ultimately, however, we can never *directly* know the mind itself. We must therefore have recourse to abstractions, derived from inferences and built into figurative models: in a word, metapsychology (see S.E. XXIII, 158–159).

Similar epistemological limitations hold for the theoretical abstractions of other branches of psychology to the extent that they too attempt to describe the inner workings of (any aspect of) the mind. This applies to even highly developed theories such as, for example, dual-route reading models (Coltheart et al. 1993), models of multiple memory systems (Schacter 1996; Schacter et al. 1998), models of divergent visual systems engaged in perception and action (Milner and Goodale 1993), and so on. *All* of theoretical psychology is ultimately just model-building of one sort or another. It is only the scale of Freud's metapsychology that distinguishes it in this respect, from the more narrowly focussed models of cognitive psychology and neuroscience. It is also (partly) for this reason that Freud's metapsychology lacks some of the specificity of modern cognitive models. But that has no bearing on their ultimate epistemological limitations.

In *Some elementary lessons in psychoanalysis*, Freud not only argued that the mind is epistemologically unknowable, but also that it is ontologically no different to the rest of nature (S.E. XXIII, 283). Kant's view was that *everything* in the world as we know it, including the contents of our external awareness, is only an indirect representation of reality. What scientists do is probe beyond this perceptual data to try to get a better picture of what Freud in his *Outline of psychoanalysis* called "the real state of affairs" (S.E. XXIII, 196). This approach, we note, is common to *all* the natural sciences—often with the use of artificial perceptual aids such as microscopes and telescopes and spectroscopy machines. They are ultimately all reduced to building *models* of our natural universe, and in this way, the mind in itself exists on the same ontological plane as the rest of nature; it is just one of the things that we perceive.

It is unquestionably significant that selective evolutionary pressures advantage organisms that develop better, that is, more accurate models of reality. In a world without vision, the first animals to evolve organs of sight would be highly advantaged. Those that develop *better* vision, for example, binocular viewing capabilities, a lens with adjustable focus, low light detection capacities for twilight conditions, etc., are further advantaged (Dawkins 1998). Moreover, those organisms that develop multiple sensory organs, each probing and sampling (and ultimately *representing*) a different aspect of the world around them are still more favoured. Considered across evolutionary time, organisms have, on this basis, developed successively better perceptually derived models of reality. Thus, the human mental apparatus (if functioning normally) delivers remarkably effective capabilities for perceptually guided locomotion, action, navigation, attentional selection, object identification and object recognition. However, the fact that the perceptual systems offer only *representations* of the world can readily be demonstrated by the remarkable errors seen in visual illusions, as well as in psychotic hallucinations and dreams.

Freud (1938) argued that the model-building of physics is no different in principle to what we do in psychoanalysis: We begin with perceptions of our inner state and then we make inferences about the true nature of the things that determine those perceptions. Our phenomenal consciousness gives us the *impression* that things are (from an external perspective) visual or auditory, or that things make us (from the internal perspective) sad or hungry, but these things are all merely *qualities* of consciousness. Our science, like all others, then strives to abstract 'the real state of affairs' that lies behind them. Freud formalised all of this in his conceptualization to the effect that consciousness has both internal and external 'perceptual surfaces' (Solms 1997a; Solms and Turnbull 2002, 18–31). The difference between psychoanalysis and the *physical* sciences is merely the perceptual surface that we use.

Behind *both* perceptual surfaces lies something else ('reality itself'), of which we can only build abstract models. Forming better models of reality itself constitutes the goal of all science, including psychoanalytic science. This may come as a surprise to those who have forgotten the origins of psychoanalysis, but for Freud his discipline had *always* been a natural science, identical *in principle* with the other basic sciences of physical reality, such as physics and chemistry. The mind in itself is therefore not ontologically different from, and not distinct from, the rest of the universe. In sum, Freud was a monist from 1900 all the way through to 1939. But his philosophical position can perhaps best be described as that of a dual-aspect monist (Solms and Turnbull 2002, 56–58) because he recognised, in *Some elementary lessons in psychoanalysis*, that "[t]he psychical, whatever its nature may be, is in itself unconscious and probably similar in kind to all other natural processes of which we have obtained knowledge" (S.E. XXIII, 283). He was also a follower of Spinoza (cf. Damasio 2004), and, indeed, in his correspondence he speaks highly of Spinoza (see Damasio 2004, 260 for an accessible survey), while in his published work he regularly describes his position in Kantian terms (see Solms 1997a, 687–689).

If the mind, in itself, is unknowable, and we can only describe it with abstract models, such as Freud's model of the 'mental apparatus' [*Seelenapparat*], then we must take full advantage of the fact that our mental apparatus can be perceived in *two different ways*. If we look at it with our eyes (via the external perceptual surface), we see a *brain*: wet, gelatinous, lobular, and embedded within the other tissues of the body. If we observe it with our internally directed perceptual surface, introspectively, we observe mental states such as thirst and pleasure. If we accept this philosophical approach, it naturally follows that we would want to make use of *both* points of view when it comes to our object of study. Why would we want to exclude, *a priori*, a full half of what we can learn about the part of nature that we are studying? In psychoanalysis, we adopt the viewpoint of subjectivity because there are things that one can learn about the nature of the mental apparatus from this perspective, which one can *never* see with one's eyes, no matter how much scientific instruments might aid them. The philosophical position taken by some other scientists (see Solms 1997a for the opinions of Francis Crick, Daniel Dennett and Gerald Edelman, for example) *does* exclude this subjective perspective. But feelings exist. They are no less real than sights and sounds and represent a fundamental part of the mind. They can also teach us a great deal about how it works. To exclude them *tout court* makes no sense at all.

When we study the mental apparatus in its physical manifestation (i.e., the brain), the information we can glean with our external sense organs is, of course, no less important. From a scientific point of view, there are a great many advantages connected with the study of physical objects. Some of our psychoanalytic colleagues (e.g. Blass and Carmeli 2007) hold a contrary and exclusionary position that we struggle to understand, not least because it seems irrational to deny oneself a source of useful data. Moreover, we ought to remind ourselves that the singular, fleeting and fugitive nature of conscious states bestows distinct disadvantages; the more stable properties of the physical brain are more amenable to the requirements of scientific method. Nevertheless, we reiterate that if one *correlates* subjective experiences with the 'wetware' of neurobiology, we are in a much stronger position to develop an accurate model of the mental apparatus itself. Thus, as with the moral of the blind men and the elephant, viewpoint-dependent errors are minimised. Neuroscience offers a second perspective on the unknowable 'thing' that we call the mental apparatus, the thing that Freud attempted to describe for the first time in his metapsychology.

To be sure, some people in the field of psychoanalysis have become anxious about how they might need to change their theories and perhaps even their practice due to advances in knowledge that stem from these neuropsychoanalytic correlations. As the saying goes, there are none so blind as those who will not see. Paradoxically, however, the interest for us personally has always gone more in the *opposite* direction. In our early careers as neuroscientists, we became frustrated with how little we were able to learn about the essential nature of the mental apparatus and the lived life of the mind with methods and theories from cognitive neuroscience that were available to us in the early 1980s. At that time, neuroscience appeared to be blind to the fact that the brain was also a sentient being capable of experiencing itself with feelings, will power and a sense of agency. The fact that these brain 'mechanisms' are endogenously driven and motivated, that they arise out of the embodied nature of the subject, substantially affects the way the apparatus operates. We feel that these are not epiphenomena, or details, or nice-to-haves, they are fundamental characteristics of how the brain works; they are what distinguishes the brain from the lung.

2.3 Scientific Foundations of Neuropsychoanalysis

The empirical basis for the discipline logically follows from the facts described in the first section of this article and from the fact that Freud lacked confidence that neuroscience would be capable of responding to questions posed by psychoanalysis in his time.

What has changed in the last few dozen years? First, there have been many technical and methodological advances in neuroscience that we reviewed here already. These in turn led to major advances in our *understanding* of the mind and its workings, most notably in the wake of abandoning behaviourism and the use of cognitive models. Thus, the last half-century has seen a dramatic advance in our understanding of episodic memory (Scoville and Milner 1957), visual attention (Posner et al. 1982), executive control (Shallice 1988) and visually guided action (Milner and Goodale 1993), to mention but a few relevant examples. However, as we have suggested elsewhere (Turnbull and Solms 2007, 1083–1084), these findings in *cognitive* neuroscience have limited implications for psychoanalysis. Of potentially greater importance are developments in the last two decades in the domain of *affective* neuroscience (Damasio 1994, 1999, 2011; Le Doux 1996, 2000; Panksepp 1998; Panksepp and Biven 2012; Turnbull and Solms 2007, 1084–1085). Significant advances in neuropsychology have also been very important, the outstanding example being the discovery of 'mirror neurons' (Rizzolati et al. 1999). Finally, one should not overlook the many developments in psychoanalysis itself in the last century. Probably the most important of these is the line of 'ethological' work on attachment, separation, and loss, beginning with Donald Woods Winnicott (1960), then through Harry Harlow (1958) and John Bowlby (1980), to Mary Ainsworth (Ainsworth et al. 1978) and Peter Fonagy (Fonagy et al. 1991; Fonagy and Target 1996).

Still, individual developments in either of neuropsychoanalysis' 'parent' disciplines do not *themselves* bridge the divide between the fields. There have been a number of bold attempts at such bridging through the decades. The work of Paul Schilder (2007), Mortimer Ostow (1954, 1955; see also Turnbull 2004) and Edwin Weinstein (Weinstein and Kahn 1955) serve as beacons in this regard. Unfortunately, none of these earlier attempts flourished into the full-fledged inter-discipline we enjoy today, in part, perhaps, because each of these early attempts ran into the same difficulties (of means, motive and opportunity) that Freud encountered (see Turnbull 2001 for an interview with Ostow on this topic).

In retrospect, one of the most central limitations may have been the lack of a well-developed dynamic *neuropsychology*. As Freud wrote, both as a neurologist and a psychoanalyst, what the localizationist neuropsychology of his time lacked was any understanding of the dynamic nature of the mental process (see Solms and Saling 1986). This only fully emerged in the 1970s, especially through the efforts of Luria (1966, 1973; see Kaplan-Solms and Solms 2000, 26–43 or Solms and Turnbull 2002, 25–27 for review). The second transformational shift occurred with the development of affective neuroscience in the 1990s (Panksepp 1998; Damasio 1994, 1999), which finally aligned neuroscience with topics that are of fundamental interest to psychoanalysis. This change allowed the disciplines to share findings not merely in relation to cognition but also in the core psychodynamic domains of emotion and instinctual drive.

The bridge-building work that catalysed our own present work in neuropsychoanalysis began with us conducting relatively conventional psychoanalytic investigations on neurological patients (Kaplan-Solms and Solms 2000). Why did this prove to be such a seminal approach for neuropsychoanalysis? In the first place, it involved a clinical method, which picked up directly from where Freud left off. The method requires relatively modest changes in working practice, and little additional training on the part of a psychoanalyst, and yet it gives direct access to the subjective mental life of the (neurological) patient in precisely the same way that psychoanalysts traditionally gather data about psychiatric (or 'normal') patients. This ensures that we can make direct observations concerning the neural correlates of metapsychological concepts in a methodologically valid setting. All of our metapsychological concepts and theories about the structure and functions of the mind are operationalised in a *clinical* psychoanalytical setting. Analytic work with neurological patients is therefore an ideal way of ensuring that we are studying the same 'things' that Freud studied, albeit from a neurological perspective.

We would like to add a further reason why clinical work in neuropsychoanalysis is best performed with *neurological*, rather than *psychiatric* patients. This is due to the methodological advantage of working with patients with *focal* brain lesions. Firstly, such patients are pre-morbidly 'typical' examples of humanity. As a population, they have none of the potentially confounding issues of aberrant development attached to psychiatric disorders (Bentall 2003, 2009). Secondly and most importantly, it enables us to correlate our psychoanalytic inferences with *definite* neuroscientific ones. Structural neurological lesions provide infinitely more precision than do psychopharmacological manipulations, given all the interactive vagaries of neurotransmitter dynamics. Moreover, by virtue of advances in structural imaging, it is possible to identify the neural basis of the clinically observed phenomena in neurological patients with a high level of scientific accuracy—a method well-suited for establishing clinico-anatomical correlations (Heilman and Valenstein 1979; Kertesz 1983; Kolb and Whishaw 1990; Lezak et al. 2004). In sum, having researched small populations of neurological patients (Kaplan-Solms and Solms 2000), we have developed a method that offers a respectable degree of experimental control, a reasonable degree of neuroanatomical localization, excellent construct validity and a direct observational window into the subjective life of the brain in a fairly naturalistic setting.

On the basis of this approach, we have been able to build a preliminary picture of how our most basic metapsychological concepts might be correlated with brain anatomy and with all that we know on the functional organisation of the brain. To take one example, in the book *Clinical Studies in Neuro-Psychoanalysis* (Kaplan-Solms and Solms 2000) we describe psychoanalytic observations on a small series of patients with right parietal lesions. They exhibited a remarkable degree of self-deception, in that they were paralyzed (on the left side) but insisted that they were *not* paralyzed. In some cases, they explained away their paralysis through transparent rationalisations ("I tired the arm out this morning doing exercises"), or they developed more complex delusions, such as claiming that the paralyzed arm does not belong to them but to the examiner or to a close relative (for examples see Aglioti et al. 1996; Feinberg 2001; Ramachandran and Blakeslee 1998). Cognitive neuroscientists have traditionally explained these remarkable clinical phenomena in terms of simple cognitive *deficits* (i.e., damage) to inferred cognitive 'modules' (for review, see Nardone et al. 2007; Turnbull et al. 2002, 2005). When we studied these patients psychoanalytically, however, we observed a pattern of psychological phenomena that was not at all modular in nature, and which was not by any means accurately defined as 'deficit'. What we observed were dynamic interactions, in which the primary forces clearly revolved around *emotional* states. Moreover, these emotionally determined dynamics rendered important aspects of the cognitive processes *unconscious*. By psychoanalytically intervening in these dynamics, it was possible to reverse the dynamic process in question and return the repressed cognitions to consciousness. This empirically demonstrated the validity of our conclusions and required students of this clinical phenomenon to radically reconceptualise its nature. As Ramachandran once humourously said: 'Of course anosognosia is a defence; it is just that neuroscientists were in denial about it'!

Our conclusion was that self-deception in right parietal lobe damage was attributable to narcissistic defensive organisations, such that the patients avoided depressive affects by using a range of primitive defence mechanisms. This regression to narcissism appeared to be attributable to a loss of capacity for whole object relationships (Kaplan-Solms and Solms 2000, 148–199). These patients also appeared to have disrupted cognitive processes that nevertheless represent space correctly, as acquired through normal development. These findings therefore confirmed the relationship between realistic spatial representations (of self/object boundaries) and the maturation of object relationships. It also established the neural correlate of what we call 'whole object' representation, the metapsychological foundation of mature object love.

However, while this approach of applying clinical psychoanalytic methods to the study of neurological patients has many strengths, it also has limitations. Because clinical observations necessarily involve limited experimental control and are open to confirmation bias (Kahneman 2003), it is a relatively weak method for determining the precise causal mechanisms involved. Experimental studies based on these purely clinical observations were thus employed to provide fuller empirical support and refinement of the above hypotheses. A series of publications (Fotopolou et al. 2004, 2008a, b; Nardone et al. 2007; Tondowski et al. 2007; Turnbull et al. 2002, 2005) have now conclusively demonstrated the powerful influence of emotions and unconscious cognitions (and associated defensive processes) in the neurodynamics that underpin the false beliefs of right parietal patients. These lines of work have been an important contribution to behavioural neurology, taking forward the ideas generated in neuropsychoanalysis beyond our own sphere or interest. As a result of these efforts, a psychoanalytical point of view is now included in conceptualisations of these phenomena in mainstream neuroscientific journals, and the influence and contribution of psychoanalysis to the neurosciences are spreading, apparently for the first time in history.

Simultaneously, psychoanalytical observations on how the mind is altered by damage to different parts of the brain has enabled us to begin to build up a coherent model of how the mental apparatus, as we understand it in psychoanalysis, is manifested in anatomy and physiology, providing what we might call a new 'physical' point of view in psychoanalytic metapsychology. By using multiple converging methods, we have made especially remarkable progress with regard to the psychoanalytic theory on dreams (Solms 1997b, 2000, 2011). For example, we have demonstrated that dreams are motivated by instinctual mechanisms and appear to protect sleep (see Solms 2001). It has been gratifying to rediscover the Freudian conception of dreams in the neurodynamics of the sleeping brain. Indeed, in 2006, at the 'Science of Consciousness' Conference in Tucson, Arizona, a formal Oxford-Rules debate (Solms vs. Hobson) on the contemporary scientific validity of the Freudian conception of dreaming resulted in a two to one vote in our favour. While such renewed demonstrations of confidence in our most basic theoretical propositions may be regarded as merely sociological phenomena, they are significant for the future viability of our discipline.

2.4 What Neuropsychoanalysis Is Not

Up to now we have described what neuropsychoanalysis *is* in terms of its historical foundations, philosophical premises and empirical underpinnings. We turn now to what neuropsychoanalysis is *not* by defining some boundary conditions.

The first boundary is a methodological one. We have especially recommended the clinico-anatomical method of making *direct* psychoanalytical observations on patients with focal brain lesions in a clinical setting. However, this is just a starting point. We have pioneered an example of how such clinical observations can be extended by using experimental neuropsychological tools. We have also already alluded to the multiple converging methods that were used to establish the neural organisation of dream psychodynamics. But numerous other approaches are possible. Thus, to take a relatively extreme instance, one might manipulate different neuropeptides in research participants who are *themselves* psychoanalysts and then have them describe their subjective states, using their expertise in doing so (i.e., with reference to the theoretical concepts that we use). Approaches such as this are rather radical, but have huge potential, and appear to be remarkably underappreciated. To take a less radical example, why do we not have systematic psychoanalytical studies on the manipulations of the different classical neurotransmitters that psychopharmacologists regularly tinker with in conventional psychiatric settings (c.f. Kline 1959; Ostow 1962, 1980; Ostow and Kline 1959)?

Other psychoanalytically informed neuroscience comes from the use of neuroimaging methods, for example, for the study of Freud's theory of mourning (Freed et al. 2009), psychodynamic aspects of confabulation (Fotopolou et al. 2004, 2008a, b; Turnbull et al. 2004a, b) or tests of Freud's dream theory (Solms 1997b, 2000). We might wonder, of course, whether work of this sort can legitimately be called 'neuropsychoanalysis', given that the data collection occurs using merely neuroscientific and psychological, rather than psychoanalytic, methods. Such work might best be described as *psychoanalytically informed neuroscience*. But who cares? On balance, we prefer to take the 'broad church' approach to this issue, such that neuropsychoanalysis represents *all* work that lies along the psychoanalysis/neuroscience boundary; it may at times involve psychoanalytically inspired neuroscience (which uses purely neuroscientific methods to test psychoanalytically informed hypotheses) and at other times the direct psychoanalytical investigation of neurological variables (brain injury, pharmacological probes, deep-brain stimulation, etc.). What unites these approaches is that they are attempts to do neuropsychoanalytic *research*, as opposed to another way of doing 'neuropsychoanalysis' that relies entirely on speculative imaginings, transpositions and guesses. The classic instances of this arise from psychoanalysts who read something about the latest developments in the neurosciences and observe that the new findings are vaguely reminiscent of such and such a phenomenon or theory in psychoanalysis. They then claim that this or that neuroscientific finding discloses the biological correlate or underpinning of some aspect of psychoanalytic theory. In our view 'armchair' speculation such as this does not represent the way forward for our field. The last century saw more than enough speculation in psychoanalysis, which resulted in the formation of multiple 'schools of wisdom' but remarkably little scientific progress. There is only one way to decide between theories and that is to *test* them against reality in such a way that the alternative predictions can be either confirmed or refuted. His *Project for a Scientific Psychology/Entwurf einer Psychologie* in 1895 was a notable early instance of such speculative guesswork, which is why Freud himself so strongly resisted its publication, describing it as an 'aberration' [*Abirrung*] (S.E. I).

One further instance of what neuropsychoanalysis is *not* deserves mention here. Neuropsychoanalysis not, in our opinion, a 'school' of psychoanalysis in the way that we currently speak of Freudian, Kleinian, Intersubjective and Self Psychology schools. Neuropsychoanalysis, we feel, is far better conceptualised as a link between *all* of psychoanalysis and the neurosciences. Otherwise, it might be described as an attempt to insert psychoanalysis into the neurosciences, as a member of the family of neurosciences, namely, the one that studies the mental apparatus from the *subjective* point of view.

Lastly, we would like to make it clear that neuropsychoanalysis (or neuroscience in general) is not a final 'court of appeal' for psychoanalysis. Psychoanalysis cannot look to any other science to tell it what errors it may have made in its methods, theory and practice. This is not to say that neuroscience brings no information to bear on what may have been erroneous or misleading paths in psychoanalysis. To take one powerful example, there is abundant evidence in neurobiology for the existence of what we refer to as 'drives' (Panksepp 1998; Pfaff 1999; Rolls 1999). For some students of psychoanalysis, drive theory has been rejected as outmoded and inappropriate (Kohut 2009; Siegel 1996). Do these neuroscientific observations invalidate this conclusion in psychoanalysis? They may not, but they are highly relevant. It may be that the term 'drive' is used in a quite different way by the psychoanalytic and neuroscientific communities (Fotopoulou et al. 2012). Or it may be that the concept of drives is more relevant to some aspects of mental life than others. Or perhaps it may be that it is only the psychoanalytic *taxonomy* of the drives that needs revision. Other interpretations are also possible. Nevertheless, it seems appropriate that the psychoanalytic community looks at the data again that led it to reject Freudian drive theory and investigates whether drives may play a more substantial part in mental life than it had previously thought. But this is not the whole story; it is merely the beginning. Once we have started to ask ourselves these questions, based on our reading of the current state of drive theory in neurobiology, we must *test* their conclusions using our own *psychoanalytic* techniques. This is bound to lead to new observations, not only of psychodynamic phenomena or continuities that we had not noticed before but also of possible limitations or errors in the neuroscientific conceptions at issue. It is, after all, more than possible that behavioural neuroscientists might have missed something important about the drives, deprived as they are of almost all the data of subjective experience.

Thus, in our opinion, the interface between psychoanalysis and neuroscience is a rather *dialectical* one. As analysts, we may learn something new about the brain that seems relevant to psychoanalysis. We may think about it, keep it at the back of our minds, entertain the possibility, but above all we *test it* psychoanalytically as well as investigate its clinical usefulness. In this way, the final court of appeal for psychoanalysis remains the psychoanalytic setting—psychoanalytic observations made on real human beings in the conventional clinical situation. A similar argument might, in principle, apply to the neurosciences, though, of course, they should and would never look to psychoanalysis as their final court of appeal. The risk of reductionism always seems to go in the direction of the physical, which is itself an interesting neuropsychoanalytical phenomenon. But neuroscientists today are looking to psychoanalysis for interesting observations and theories, which they are increasingly applying to their work. They also quite naturally adopt them where they seem appropriate (Feinberg 2001; Fotopolou et al. 2004, 2008a, b; Ramachandran and Blakeslee 1998; Turnbull et al. 2002, 2004a, b) and then move on.

2.5 The Future

In the sciences there is a long history of remarkable creativity at the boundaries between disciplines as for example in molecular biology (Watson and Crick 1953) or in cognitive neuroscience (Bowman and Turnbull 2009). Consistent with this, our interdisciplinary field has already opened rich veins of new enquiry. Doubtless this will continue to occur and do so in unpredictable ways. Nevertheless we would like to sketch a general outline of how we hope the field will develop.

Our vision involves the collaborative investigation of phenomena that have a common interest and are approached using a level of rigour associated with all good scientific enquiry. But these investigations must also respect the methodological tools (with all the advantages and disadvantages) associated with each distinct field. It would be a shame if neuropsychoanalysis were to become an armchair activity or a field based on speculation rather than empirical work, a field in which the acquisition of knowledge is unidirectional (i.e., neuroscience only informs psychoanalysis and not *vice versa*), a discipline that studies scientific phenomena that are only *nominally* psychoanalytic but lacks the deep respect for the subjective perspective, which is the hallmark of psychoanalysis.

We are confident that this will not happen. For, as Freud told Einstein 83 years ago, "There is no greater, richer, more mysterious subject, worthy of every effort of the human intellect [*menschlichen Intellekts*], than the life of the mind [*das Seelenleben*]" (Freud 1929, 110).

References

Aglioti, S., Smania, N., Manfredi, M., & Berlucchi, G. (1996). Disownership of the left hand and objects related to it in a right brain damaged patient. *NeuroReport, 8*, 293–296.
Ainsworth, M. D. S., Blehar, M. C., Waters, E., & Wall, S. (1978). *Patterns of attachment: A psychological study of the strange situation*. Hillsdale: Erlbaum.
Bentall, R. (2003). *Madness explained: Psychosis and human nature*. London: Penguin Books Ltd.
Bentall, R. (2009). *Doctoring the mind*. London: Allen Lane.
Berger, H. (1929). Über das Electrenkephalogramm des Menschen. *Archives für Psychiatrie Nervenkrankheiten, 87*, 527–570.
Blass, R., & Carmeli, Z. (2007). The case against neuropsychoanalysis: On fallacies underlying psychoanalysis' latest scientific trend and its negative impact on psychoanalytic discourse. *The International Journal of Psychoanalysis, 88*(1), 19–40.
Bowlby, J. (1980). *Attachment and Loss*. London: Penguin.
Bowman, C., & Turnbull, O. H. (2009). Schizotypy and flexible learning: A pre-requisite of creativity. *Philoctetes, 2*, 5–30.
British Psychoanalytic Society. (2008). *English-speaking conference debate*. London: British Psychoanalytic Society.
Coltheart, M., Curtis, B., Atkins, P., & Haller, M. (1993). Models of reading aloud: Dual-route and parallel-distributed-processing approaches. *Psychological Review, 100*, 589–608.
Damasio, A. (1994). *Descartes error: Emotion, reason, and the human brain*. London: Picador.
Damasio, A. (1999). *The feeling of what happens: Body and emotion in the making of consciousness*. London: William Heinemann.

Damasio, A. (2004). *Looking for Spinoza*. London: Vintage.
Damasio, A. (2011). *Self comes to mind: Constructing the conscious brain*. London: William Heinemann.
Dawkins, R. (1998). *Unweaving the rainbow*. London: Penguin.
Feinberg, T. E. (2001). *Altered egos: How the brain creates the self*. Oxford: Oxford University Press.
Finger, S. (1994). *Origins of neuroscience: A history of explorations into brain function*. New York: Oxford University Press.
Fonagy, P., & Target, M. (1996). Playing with reality: I. Theory of mind and the normal development of psychic reality. *International Journal of Psycho-Analysis, 77*, 217–233.
Fonagy, P., Steele, H., & Steele, M. (1991). Maternal representations of attachment during pregnancy predict the organization of infant-mother attachment at one year of age. *Child Development, 62*, 891–905.
Fotopolou, A., Solms, M., & Turnbull, O. H. (2004). Wishful reality distortions in confabulation. *Neuropsychologia, 42*, 727–744.
Fotopoulou, A., Conway, M. A., Solms, M., Tyrer, S., & Kopelman, M. (2008a). Self-serving confabulation in prose recall. *Neuropsychologia, 46*, 1429–1441.
Fotopoulou, A., Conway, M. A., Tyrer, S., Birchall, D., Griffiths, P., & Solms, M. (2008b). Is the content of confabulation positive? An experimental study. *Cortex, 44*, 764–772.
Fotopoulou, A., Pfaff, D., & Martin, C. (Eds.). (2012). *From the couch to the lab: Trends in psychodynamic neuroscience*. Oxford: Oxford University Press.
Freed, P. J., Yanagihara, T. K., Hirsch, J., & Mann, J. J. (2009). Neural mechanisms of grief regulation. *Biological Psychiatry, 66*(1), 33–40.
Freud, S. (1891). *On Aphasia: A critical study*. London: Imago.
Freud, S. (1895). *Project for a scientific psychology*. S.E. I, 281–397.
Freud, S. (1914). *On narcissism: An introduction*. S.E. XIV, 67–102.
Freud, S. (1915). *The unconscious*. S.E. XIV, 159–215.
Freud, S. (1920). *Beyond the pleasure principle*. S.E. XVIII, 1–64.
Freud, S. (1929). Letter to Einstein, 1929. In Ilse Grubrich-Simitis (1995), 'No greater, richer, more mysterious subject [...] than the life of the mind'. An early exchange of letters between Freud and Einstein. *International Journal of Psychoanalysis, 76*, 115–122.
Freud, S. (1938). *Some elementary lessons in psychoanalysis*. S.E. XXIII, 279–286.
Freud, S. (1940). *An outline of psychoanalysis*. S.E. XXIII, 139–207.
Harlow, H. (1958). The nature of love. *American Psychologist, 13*, 673–685.
Heilman, K. M., & Valenstein, E. (1979). *Clinical neuropsychology*. Oxford: Oxford University Press.
Kahneman, D. (2003). A perspective on judgement and choice. *American Psychologist, 58*, 697–720.
Kaplan-Solms, K., & Solms, M. (2000). *Clinical studies in neuro-psychoanalysis: Introduction to a depth neuropsychology*. New York: Karnac Books.
Kertesz, A. (1983). *Localisation in neuropsychology*. New York: Academic.
Kline, N. S. (1959). *Major problems and needs in psychopharmacology frontiers*. Boston: Little Brown.
Kohut, H. (2009). *The restoration of the self*. Chicago: Chicago University Press.
Kolb, B., & Whishaw, I. Q. (1990). *Fundamentals of human neuropsychology*. New York: Freeman & Co.
Le Doux, J. (1996). *The emotional brain*. New York: Touchstone.
Le Doux, J. (2000). Emotion circuits in the brain. *Annual Review of Neuroscience, 23*, 155–184.
Lezak, M. D., Howieson, D. B., & Loring, D. W. (2004). *Neuropsychological assessment* (4th ed.). New York: Oxford University Press.
Luck, S. J. (2005). *An introduction to the event-related potential technique*. Cambridge, MA: MIT Press.
Luria, A. R. (1966). *Higher cortical function in man*. New York: Basic Books.
Luria, A. R. (1973). *The working brain*. Aylesbury: Penguin.
Mayberg, H., Lozano, A., Voon, V., McNeely, H. E., Seminowicz, D., Hamani, C., Schwalb, J. M., & Kennedy, S. H. (2005). Deep brain stimulation for treatment-resistant depression. *Neuron, 45*, 651–660.

Milner, A. D., & Goodale, M. A. (1993). Visual pathways to perception and action. *Progress in Brain Research, 95*, 317–337.

Nardone, I. B., Ward, R., Fotopoulou, A., & Turnbull, O. H. (2007). Attention and emotion in anosognosia: Evidence of implicit awareness and repression? *Neurocase, 13*(5), 438–445.

Ostow, M. (1954). A psychoanalytic contribution to the study of brain function. 1: The frontal lobes. *Psychoanalytic Quarterly, 23*, 317–338.

Ostow, M. (1955). A psychoanalytic contribution to the study of brain function. 2: The temporal lobes. 3: Synthesis. *Psychoanalytic Quarterly, 24*, 383–423.

Ostow, M. (1962). *Drugs in psychoanalysis and psychotherapy.* New York: Basic Books.

Ostow, M. (1980). *The psychodynamic approach to drug therapy.* New York: Van Nostrand Reinhold.

Ostow, M., & Kline, N. S. (1959). The psychic actions of reserpine and chlorpromazine in psychopharmacology frontiers. In N. S. Kline (Ed.), *Major problems and needs in psychopharmacology frontiers* (pp. 45–58). Boston: Little Brown.

Panksepp, J. (1998). *Affective neuroscience: The foundations of human and animal emotions.* Oxford: Oxford University Press.

Panksepp, J., & Biven, L. (2012). *The archeology of mind.* New York: W. W. Norton.

Penfield, W. (1952). Memory mechanisms. *Archives of Neurology & Psychiatry, 67*, 178–198.

Pfaff, D. W. (1999). *Drive: Neurobiological and molecular mechanisms of sexual motivation.* Cambridge, MA: MIT Press.

Posner, M. I., Cohen, Y., & Rafal, R. D. (1982). Neural systems control of spatial orienting. *Philosophical Transactions of the Royal Society of London Series B-Biological Sciences, 298*, 187–198.

Ramachandran, V. S., & Blakeslee, S. (1998). *Phantoms in the brain: Human nature and the architecture of the mind.* London: Fourth Estate.

Rizzolati, G., Fadiga, L., Fogassi, L., & Gallese, V. (1999). Resonance behaviours and mirror neurons. *Archives of Italian Biology, 137*, 85–100.

Rolls, E. T. (1999). *The brain and emotion.* Oxford: Oxford University Press.

Schacter, D. L. (1996). *Searching for memory.* New York: Basic books.

Schacter, D. L., Norman, K. A., & Koutstaal, W. (1998). The cognitive neuroscience of memory. *Annual Review of Psychology, 49*, 289–318.

Schilder, P. (2007). *Brain and personality: Studies in the psychological aspects of cerebral neuropathology and the neuropsychiatric aspects of the motility of schizophrenics.* Whitefish: Kessinger Publishing.

Scoville, W. B., & Milner, B. (1957). Loss of recent memory after bilateral hippocampal lesions. *Journal of Neurology, Neurosurgery, and Psychiatry, 20*, 11–21.

Shallice, T. (1988). *From neuropsychology to mental structure.* Cambridge: Cambridge University Press.

Siegel, A. M. (1996). *Heinz Kohut and the psychology of the self.* London: Routledge.

Solms, M. (1997a). What is consciousness? *Journal of the American Psychoanalytic Association, 45*, 681–703.

Solms, M. (1997b). *The neuropsychology of dreams.* Mawah: Lawrence Earlbaum Press.

Solms, M. (1998). Before and after Freud's "Project". *Neuroscience of the Mind on the Centennial of Freud's Project for a Scientific Psychology. Annals of the New York Academy of Sciences, 843*, 1–10.

Solms, M. (2000). Dreaming and REM sleep are controlled by different brain mechanisms. *Behavioral and Brain Sciences, 23*, 843–850.

Solms, M. (2001). The interpretation of dreams and the neurosciences. *Psychoanalysis and History, 3*, 79–91.

Solms, M. (2002). An introduction to the neuroscientific works of Sigmund Freud. In G. van de Vijver & F. Geerardyn (Eds.), *The pre-psychoanalytic writings of Sigmund Freud* (pp. 25–26). London/New York: KarnacBooks.

Solms, M. (2011). Neurobiology and the neurological basis of dreaming. In P. Montagna & S. Chokroverty (Eds.), *Handbook of clinical neurology, 98 (3rd Series), sleep disorders – Part 1* (pp. 519–544). New York: Elsevier.

Solms, M., & Saling, M. (1986). On psychoanalysis and neuroscience: Freud's attitude to the localizationist tradition. *The International Journal of Psycho-Analysis, 67*, 397–416.

Solms, M., & Turnbull, O. H. (2002). *The brain and the inner world: An introduction to the neuroscience of subjective experience.* New York: Other Press/Karnac Books.

Sulloway, F. J. (1979). *Freud: Biologist of the mind.* Bungay: Chaucer Press.

Sutton, S., Braren, M., Zubin, J., & John, E. R. (1965). Evoked-potential correlates of stimulus uncertainty. *Science, 150*(3700), 1187–1188.

Sutton, S., Tueting, P., Zubin, J., & John, E. R. (1967). Information delivery and the sensory evoked potential. *Science, 155*(3768), 1436–1439.

Tondowski, M., Kovacs, Z., Morin, C., & Turnbull, O. H. (2007). Hemispheric asymmetry and the diversity of emotional experience in anosognosia. *Neuropsychoanalysis, 9*, 67–81.

Turnbull, O. H. (2001). The neuropsychology that would have interested Freud most. *Neuro-Psychoanalysis, 3*(1), 33–38.

Turnbull, O. H. (2004). Founders of neuro-psychoanalysis: Interview with Mortimer Ostow. *Neuro-Psychoanalysis, 6*(2), 209–216.

Turnbull, O., & Solms, M. (2007). Awareness, desire, and false beliefs: Freud in the light of modern neuropsychology. *Cortex, 43*, 1083–1090.

Turnbull, O. H., Berry, H., & Evans, C. E. Y. (2004a). A positive emotional bias in confabulatory false beliefs about place. *Brain & Cognition, 55*, 490–494.

Turnbull, O. H., Jenkins, S., & Rowley, M. L. (2004b). The pleasantness of false beliefs: An emotion-based account of confabulations. *Neuro-Psychoanalysis, 6*(1), 5–16.

Turnbull, O. H., Jones, K., & Reed-Screen, J. (2002). Implicit awareness of deficit in anosognosia: An emotion-based account of denial of deficit. *Neuropsychoanalysis, 4*, 69–86.

Turnbull, O. H., Owen, V., & Evans, C. E. Y. (2005). Negative emotions in anosognosia. *Cortex, 41*, 67–75.

Walter, W. G., Cooper, R., Aldridge, V. J., McCallum, W. C., & Winter, A. L. (1964). Contingent negative variation: An electric sign of sensorimotor association and expectancy in the human brain. *Nature, 203*, 320–384.

Watson, J. D., & Crick, F. H. C. (1953). Molecular structure of nucleic acids: A structure for deoxyribose nucleic acid. *Nature, 171*, 737–738.

Weinstein, E. A., & Kahn, R. L. (1955). *Denial of illness: Symbolic and physiological aspects.* Springfield: Charles C. Thomas.

Winnicott, D. W. (1960). The theory of the parent-infant relationship. *International Journal of Psychoanalysis, 41*, 585–595.

Part II
Embodiment as Bridge Between Psychoanalysis and Neuroscience

Chapter 3
Enactments in Transference: Embodiment, Trauma and Depression. What Have Psychoanalysis and the Neurosciences to Offer to Each Other

Marianne Leuzinger-Bohleber

Abstract The current controversy within psychoanalysis whether or not 'historical truth' plays a role in successful patient treatment is fuelled by neuroscientific research in the field of Embodied Cognitive Science. The insights from the neurosciences provide an objective aspect to the subjective, 'narrative' aspect so important in psychoanalytical therapy. The author concludes that memory is always based on new and idiosyncratic (hermeneutic) narratives that take place in interactional situations in the present, which nevertheless contain traces of historical truth. Leuzinger-Bohleber regards neuroscientific findings of neuroscience as coherent with those from psychoanalysis, and, therefore, stresses the importance of prevention and early intervention, especially in light of new epigenetic research.

Keywords Patient treatment • Embodied Cognitive Science • Enactment • Trauma • Sexual abuse • Memory • Epigenetics

3.1 Countertransference Reactions as a Product of Unconscious Enactments or Embodied Memories of Trauma?

Before I can properly open the door for Ms. M., she rushes into my office. She grabs my hand taking it between both of hers and holds it tight in a strange 'sexually stimulating' way. At the same time she comes very close to me, overstepping the normal boundaries of bodily proximity: "Hellöchen,[1] I am so, so happy that I can talk to you…" I immediately become aware of a strong negative countertransference reaction connected with intense tension in my stomach and other averse bodily

---

[1] A strange and infantile way to say "hi."

M. Leuzinger-Bohleber (✉)
Sigmund-Freud-Institut, Beethovenplatz 1-3, D-60325 Frankfurt am Main, Germany
e-mail: m.leuzinger-bohleber@sigmund-freud-institut.de

© Springer International Publishing Switzerland 2016
S. Weigel, G. Scharbert (eds.), *A Neuro-Psychoanalytical Dialogue for Bridging Freud and the Neurosciences*, DOI 10.1007/978-3-319-17605-5_3

reactions. I observe some of my own thoughts: "What an overwhelming woman! I don't like this bodily contact: it is too much and not appropriate...! Why did I offer her a session? Will I ever get rid of her again...?" Then she asks me where she can find the toilet and perturbs me further by leaving the door open. Only when she finally sits down in her chair do I notice that she has a pretty, girlish face with a constant, sociable smile (according to Benecke and Krause 2001) and a beautiful female body although she is dressed like a man in simple jeans and a baggy, plain sweater. She is in her mid-forties.

Ms. M. tells me that she was sent to me by a general practitioner because of a psychosomatic and psychic breakdown: For weeks she was suffering from anxiety and could not sleep or eat properly. She suffered from headaches and diffuse heart symptoms. She went to the doctor convinced that she had a somatic illness, but he did not find anything. "I don't have any idea why I don't function anymore... I have always been a tough and perfectly functioning mother and social worker...." When I ask her for the context of her breakdown, she finally mentions that on that extraordinary day her lover of one year, who is married to someone else, told her that he is moving to another city and will not be able to visit her every Friday night anymore. Only later in analysis did we understand that her "falling into a deep, deep hole..." was probably triggered by this unexpected experience of losing a (former) love object in combination with her 14-year-old daughter's process of adolescent separation and individuation.

In reference to this case I would like to discuss the following three theses in this article:

3.1.1 Concerning Embodied Memory

The first thesis is that already in the very first interaction with Ms. M., in the "initial scene," as Hermann Argelander (1987) characterized it, she "remembered" early traumatizations but not through any sort of "verbal schemata." Instead she acted out specific unconscious fantasies through sensorimotor-affective coordinations that became observable in her bodily enactments and in my countertransference reactions, although neither of us understood this as such at that time. These clinical observations may serve as one example of how psychoanalytical research within the psychoanalytical situation might be of great interest for interdisciplinary collaborations, as neuroscientists, cognitive scientists and cultural anthropologists investigate memory in a dialogue that is fruitful for all sides. I want to illustrate this briefly with the concept of *embodied memory* that Rolf Pfeifer and I presented in several papers already back in the 1990s. It has since been taken up broadly in many fields although sometimes in a popular, superficial way (Pfeifer and Leuzinger-Bohleber 1986, 1989, 1992; Leuzinger-Bohleber and Pfeifer 2011, 2013; Leuzinger-Bohleber in press a, b).

According to embodied cognitive science, memory is not to be conceived as stored structures but as a function of the whole organism, as a complex, dynamic, recategorizing and interactive process that is always 'embodied.' It is important to note that sensorimotor, embodied coordination does not simply mean nonverbal: it implies that there is a *coupling* between the sensory and the motor processes, i.e., the two mutually influence one another. Biologically, this coupling is implemented via neural maps embedded in the sensorimotor systems of the organism. William J. Clancey (1993) thus defines memory as the ability to organize neurological processes in such a way that they coordinate and therefore categorize sensory and motor processes in a way that is similar to past situations. This conceptualization of memory is central to the discussion of a main controversy within current psychoanalysis on the role of 'historical versus narrative truth,' particularly in trauma. Peter Fonagy and Mary Target (1997, 216), in examining the results of recent memory research, have postulated: "[…] whether there is historical truth and historical reality is not our business as psychoanalysts or psychotherapists."[2]

In our view their thesis has been contradicted by the concepts, the valuable clinical research with traumatized patients and the empirical findings from the DPV follow-up study (Leuzinger-Bohleber et al. 2002) as well as the ongoing LAC depression study (Leuzinger-Bohleber et al. 2010a, b). To summarize our conclusions, which we have elaborated in detail elsewhere (Leuzinger-Bohleber and Pfeifer 2002; Leuzinger-Bohleber et al. 2008b; Leuzinger-Bohleber in press a, b), memory always consists of new and constructive processes in the 'here and now' of a current interactional situation (system—environment—interaction) which is indispensible for constituting memories. At the same time, the constitution of memories is not arbitrary. This is because of how a given system-environment interaction is structured and the way the sensorimotor patterns are interpreted and determined by an individual's history. Memories are constructed by analogy to previous situations with similar sensorimotor patterns. Although this physical stimulation is

---

[2] Fonagy and Target (1997) conclude their excellent overview "Perspectives on the recovered memories debate" with the following: "Unconscious memory is implicit memory. The psychotherapist or psychoanalyst's pressure on the patient to find the episodic roots of these memory traces is doomed to failure, as episodic experience is stored separately, without the significance for the determination of behavior, expectation, and belief that common-sense psychology attributes to it. The recovery of the episodic roots of implicit memories leads to illusory experiences, not to psychic change. Change will occur through the re-evaluation of mental models, or the understanding of self-other representations implicitly encoded as procedures in the human mind. Change is a change of form more than of content: therapy modifies procedures, ways of thinking, not thoughts. Insight or new ideas, by themselves, cannot sustain change. The internalization of this therapeutic process as an indication for appropriate termination of therapy implies a change in mental models, an alteration of the hierarchical organization of implicit memory procedures. It is not necessarily associated with increased self-awareness as a specific self-conscious activity. Recovered memory therapies are in pursuit of a false goal. There can be only psychic reality behind the recovered memory- whether there is historical truth and historical reality is not our business as psychoanalysts or psychotherapists" (215 f.). Although we sympathize with the authors' political position and agree that psychoanalysts or psychotherapists should "avoid, wherever possible, becoming entangled in legal procedures concerning CSA" (childhood sexual abuse, M.L.-B.; Fonagy and Target 1997, 209) we do not share all the conclusions from the studies mentioned above in the field of embodied cognitive science and our own clinical experiences (see also Bohleber and Leuzinger-Bohleber in press).

always subject to interpretation, which varies depending on an individual's history, the sensory stimulation itself is still 'objective' and not arbitrary. This is a consequence of embodiment: sensorimotor states are, at least theoretically, measurable physical processes; the sensorimotor coordination is established by the way the neural maps are integrated in a single organism and is thus also 'objective'. In this sense memories result from constructive processes on the one hand but are influenced by the 'historical truth' on the other hand, which means that, for example, the processes dealing with a (traumatic) situation that formed first historically speaking constrain the recategorization of the new analogous situation. Consequently, *recategorizations in later interactional situations are related to the original trauma.* Metaphorically, we could postulate that memory is always based on new and idiosyncratic (hermeneutic) narratives that take place in the present interactional situation, but at the same time memory contains traces of the 'historical truth.'

3.1.2 Concerning Trauma

My second thesis is that embodied memory[3] is particularly interesting for understanding traumatized patients because trauma, due to its extreme quality and consequences, is also inscribed in the body in an extreme way (see Bohleber 2000). As we know, the symbolization and verbalization of trauma is difficult and sometimes even impossible. Many psychoanalysts use the metaphor "Trauma is inscribed in the body." With the concept of 'embodiment' cognitive scientists (Rolf Pfeifer and his group for example) and neuroscientists (e.g., Fuchs et al. 2010) try to describe the exact same phenomena. While we observe the phenomenon in all of our patients, those patients that we consider 'healthier' are the ones who are more flexible and dynamic with their recategorizations. Traumatic experiences are defined by their extreme quality which means that *the functional recategorization and adaptation to new interactional situations is often limited, severely disturbed and sometimes even impossible*. Therefore, the trauma becomes observable in embodied memories and enactments as strange, bizarre and inadequate behaviors as with Ms. M. in her first interaction with me.

3.1.3 Concerning Treatment

Finally, for my third thesis, I would like to use clinical evidence to illustrate that severely traumatized patients, such as Ms. M., suffer from a high frequency of comorbidity (for example, PSD in combination with severe depression and

---

[3] This specific concept of embodied memory was new 10 years ago and has been taken up by many authors in cognitive neuropsychology in the meantime, although its radical dimension has been greatly reduced.

personality disorders). Therefore, short term treatment in therapy cannot help them. *Psychoanalysis* with modified techniques in treatment *is still the prescribed method of psychotherapeutic help for these patients*. At the same time, treating these patients for years also offers unique research possibilities leading to insights in the long-term effects of early traumatizations that may help explain new findings in developmental and epigenetic research on the relationship between genetic vulnerability, trauma and depression.

3.2 Clinical Psychoanalytical Research and the Functions of Psyche and the Brain: Human Interaction, Affects, Memories and Trauma in the Transference Relationship to the Analyst

In what follows I concentrate on aspects of clinical findings that might enrich an embodied or neuroscientific understanding of memory. My central claim is that Ms. M. does not enact a stored, unconscious memory of traumatic experiences with her primary object in the transference of a 'statically stored, fixed and verbal' unconscious truth: "I have to grab on quickly and blatantly to my love object demonstrating to her how grateful I am for being in contact with her—otherwise the love object abandons me, which means for me 'falling into a deep, deep hole,' a catastrophe...." Instead, she constructs these memories by coordinating current sensorimotor stimulations in the interaction with the analyst, an important 'Other' upon whom she will be existentially dependent for several years in a way similar to her dependence on her primary object from the original traumatic situation. In this manner—rushing through my office door, grasping my hand, etc.—Ms. M. 'remembers' traumatic experiences of being overwhelmed and psychologically abused by the primary object and later during adolescence of being sexually abused.

3.2.1 Remembering Psychological and Sexual Abuse

Of course, only through psychoanalysis could we begin to understand these memories. Before they were for the most part unknown (unconscious) to Ms. M. Here is a summary of the most important biographical findings:

Shortly before his traumatic experiences during incarceration in Russia Ms. M.'s 52-year-old father told her, shortly before his death when Ms. M. was in her adolescence, that he left his wife when Ms. M. was ten years old because he could not stand his spouse's bitterness and rigidity anymore. These traits of her mother were often a central issue during psychoanalysis. For a long time she protected her mother from critical observations and guarded her own unconscious knowledge of how terribly she had suffered due to her mother's lack of empathy and warmth for her as a child and how destructive the chronic psychic abuse had been. Her mother's miserable childhood as an orphan in WWI, the years during WWII and a dramatic rape in 1945 provide Ms. M. with "explanations" for her "mother's problems and deficiencies." During the third year of treatment, dreams lead to the hypothesis that she might have witnessed her mother being raped, a hypothesis which was strengthened when she asked her mother about the incident: When she was three years old she observed her mother being raped by three Russian soldiers in a very cruel and frightening way. Her mother now told her how traumatizing this event had been for her: "Since this experience I hated and detested my female body and never wanted sexuality again, perhaps one reason for the failure of our marriage," her mother said.

Ms. M. grew into the role of the 'perfect daughter,' while her brother seemed to withdraw from family life. He failed in school and immigrated to Canada when he was 18. In contrast, Ms. M. became the "the apple of her mother's eye" ("Augapfel") and tried to please her by getting good grades in school, particularly in art and music. Until she was 16 she slept in the "marriage bed" with her mother and spent her spare time and holidays almost exclusively with her. When she was 15, her uncle abused her sexually in his studio where she took art lessons. She did not resist. Only during psychoanalysis did she realize how harmful these experiences had been for her. She had felt guilty and responsible for the abuse "because I was longing for love and tenderness so much. I was not able to fight for normal boundaries." Afterwards she developed a variety of psychosomatic symptoms: migraines, sleep disturbances and bulimia. At that time she did not receive any professional help.

Despite her symptoms she managed to finish school successfully and started attending university, still living with her mother. The 1968 student revolts afforded her at least a minimal (physical) distantiation from her mother: she moved into a shared flat with other female students and had numerous sexual affairs. This behavior might be seen as being of a promiscuous nature. She was exposed several times to cruel and dangerous situations. In psychoanalysis she understood that these had been enactments in a severe state of dissociation. In this "strange state of mind" she enacted unconscious fantasies connected to the rape of her mother. Psychodynamically she seemed to "prove" her unconscious truth: "I do not deserve a better fate than my mother." She had seven abortions within ten years. Once she finished university, she chose very stressful jobs working with adolescents with drug abuse problems and in juvenile detention centers. She also worked with extremely ill cancer patients and for the past ten years now she has been working with juvenile delinquents from high-risk areas in her city.

In treatment we came to understand that the "flight into an extremely stressful 12-h work day" was an unconscious attempt to live her own life in her own apartment separate from her mother. It was also a manic defense against severe depression. Although she longed for a family of her own and had numerous love relations, she always had to break them off after a short period of time. At the age of 35, she adopted a severely handicapped girl, Anna. She moved into the same house as her mother who then took care of the child while Ms. M. was at work during the day. At the age of 38, she unexpectedly became pregnant again. "Of course" she planned another abortion. But after Anna nearly died from an asthma attack, it became clear to Ms. M. how fragile the equilibrium of her life was and how real a breakdown of that equilibrium would be if she lost Anna. Thus, she decided to give birth to the child. Raising her younger, healthy daughter Marion and Anna became the common center of her and her mother's lives. They bought a house together and established an 'open door' policy, which meant always keeping all the doors in the house open (e.g., they never closed the door to the bathroom while urinating). Their life together was characterized by harmony on the surface and a shared sense of purpose ("Lebenssinn"). They even shared Ms. M.'s lover who seemed to be as equally attached to Ms. M. as he was to her mother. They seemed to have found a kind of a stable balance until Marion entered adolescent development and the simultaneous sudden loss of the boyfriend, a "good friend of the whole family." The combination of the two provoked Ms. M.'s severe crisis and psychic breakdown.

3.2.2 Remembering and Denying Early Traumatic Separations

Other 'embodied memories' seemed to be connected to Ms. M.'s early traumatic loss of the primary object. Again, this was only understood after many, many psychoanalytic sessions:

Shortly after her birth in 1942, Ms. M.'s mother received news that her husband was missing in action on the Russian front, which was one reason for her psychic breakdown and her incapacity to care for and breast feed her baby. After three months of severe depression, she gave the baby to her mother-in-law, a "hard and staunch National Socialist." Years later this same mother-in-law would still report proudly that she "educated her granddaughters strictly following Johanna Harer's book" *Die deutsche Mutter und ihr erstes Kind* (The German Mother and Her First Child). To mention just one example, the baby was locked in the basement for two nights where her crying could not be heard: "Afterwards she never cried at night anymore." In general, the young Ms. M. developed into a strikingly well-educated and brave little girl. At the age of one, she was already clean, obedient and "easy to take care off," which were some of the reasons why her psychically instable mother dared to take her back at the end of 1944.

We finally understood in analysis that the loss of her lover was probably due to the unconscious recollection of the early traumas of living with a severely depressive primary object and finally losing her—an experience of despair and panic, "falling into a deep, deep hole." In the initial interview, she also seemed to remember that making contact with an 'important Other' whom you depend on can only be done by grabbing on to her, holding her tight and playing the grateful "sunshine child." In my countertransference fantasy of "How can I ever get rid of her," I unconsciously perceived the message that the patient had gone through an early traumatic object loss, an experience that should not be repeated in the therapeutic relationship.

3.3 Memory, Trauma and Depression and the Dialogue Between Psychoanalysis, Embodied Cognitive Science and Epigenetics

As discussed in the papers mentioned above, the 'objective' biographical information (the mother's depressive illness, early separation in the first weeks of her life, sexual abuse, etc.) proved to be helpful in Ms. M.'s psychoanalytic sessions to finally recognize the traces of these 'historical traumata' in her current behavior (e.g., seeing the similarities between her current psychosomatic and emotional reactions and those of a baby interacting with a depressive, helpless mother and then trying to be her "sunshine" in order to revitalize the "dead mother"; see for instance Stern 1995; Green 1999). Thus, we think that for stable therapeutic change in our patients both approaches are indispensible: understanding the idiosyncratic ways of unconscious functioning (see Bollas 1992; Green 1999; Hinshelwood 1991; Laub 2005; Sandler and Sandler 1997) and recategorizations as well as the attempt to understand the highly individual, biographical (historical) truth as the "specific, undeniable reality of trauma" (see also Bohleber 2000, 2005; Fischer and Riedesser 1998; Van der Kolk et al. 1996).[4]

It is fascinating that the new methods of psychiatric and epigenetic research seem to produce results that confirm the clinical psychoanalytical research on the detrimental effects of early trauma and depression as illustrated here by the clinical example of Ms. M. Jonathan Hill (2009) summarized findings from current developmental research on adult depression. Many studies showed an increase in the probability of developing depression as an adult after neglect or the loss of a parent early in life (Hill 2009, 200 ff.; Bifulco et al. 1987; Hill et al. 2001). Fergusson and Mullen (1999) reviewed the extensive literature on the role of childhood sexual abuse and showed that the association with depression in adulthood was highly robust: a history of childhood sexual abuse increased the risk of depression by a factor of approximately four.

---

[4] To repeat our thesis once again: Experiences and memories have an objective and a subjective aspect. The *objective* one is given by patterns of sensory stimulation in a particular sensory-motor interaction, which is, in principle, physically measurable. The *subjective* aspect refers to how individual experiences associated with these patterns, which is the result of the constructive process of interpretation, are determined by an individual's history. The sensory stimulation to which the organism is 'objectively' exposed is not a matter of passively undergoing physical stimulation but it is rather generated as the organism interacts with its environment. As a consequence of this interaction the resulting patterns of sensory stimulation are structured and contain correlations which can be easily interpreted by neural mechanisms. The types of interactions are in turn a result of developmental processes. The experiences of extreme pain and bodily unease and loneliness, all experiences that Ms. M. went through as a baby in her first three months, have a determining influence because the developmental processes strongly depend on the adequacy, richness and structure of sensory stimulation. This might be one of the reasons for her chronic manic defense: her constant attempts at self-stimulation and her neglecting any signs of exhaustion, tiredness, etc. until the "total breakdown" mentioned above. Thus, it is the sensorimotor coupling that provides the basis from which the developmental processes can be bootstrapped.

Other important, new insights on the *relationship between trauma and depression* are being discussed in the relatively new field of *epigenetic research*.[5] Twin studies have established that unipolar depression is moderately heritable (Kendler et al. 2006; Hill 2009, 202 ff.). Recent research in epigenetics, however, shows that even with a genetic propensity, depression results only if the person additionally goes through severe early traumatizations. Avshalom Caspi et al. (2003) were able to show that only early experiences of trauma trigger the short allele of the 5-HHT gene that regulates relevant neurotransmitters and might later lead to depression. If no such early trauma occurs, depression is not observed later. These findings are of extreme importance for many professionals, especially psychoanalysts. They support our clinical findings that early prevention and intervention in cases of depressive children, adolescents and adults, even those coming from genetically burdened families, as with Ms. M., can be helpful and effective in strengthening the resiliency of these at-risk individuals.

To return to the topic of this chapter, these contemporary epigenetic and neurobiological studies give new support to very old findings in 'classical' psychoanalysis, for example, the famous study by René Spitz on anaclitic depression and hospitalism in the 1940s that impressively showed how early separation trauma can contribute to severe depression even in infancy. As is well known, Robertson and Robertson (1975) replicated these findings in the 1970s with influential studies on early separation. At that time there was already discussion about the interesting correspondence between these clinical empirical psychoanalytical findings and those of Harry Harlow's famous experiments with monkeys in the 1950s. Thanks to modern research instruments, Steven Suomi (2010), a follower of Harlow's, was able to demonstrate that early separation trauma has an enormous influence on *neurobiological* determining factors that contribute to the development of aggression, anxiety, social integration and, consequently, to the survival of genetically predisposed Rhesus monkeys.

According to Suomi's studies the influences of early trauma are also transmitted to the next generation, a finding that again corroborates details from clinical psychoanalytical observations with traumatized patients and their families

---

[5] In 2003 Caspi and his research team published a fascinating paper in *Science*, "Influence of Life Stress on Depression: Moderation by a Polymorphism in the 5-HTT Gene." In a prospective longitudinal study on a representative birth cohort, the researchers tested why stressful experiences lead to depression in some people but not in others. A functional polymorphism in the promoter region of the serotonin transporter (5-HTT) gene was found to moderate the influence of stressful life events on depression. Individuals with one or two copies of the short allele of the 5-HTT promoter polymorphism exhibited more depressive symptoms, diagnosable depression, and suicidality in relation to stressful life events than individuals who were holozygous for the long allele. Their epidemiological study thus provided evidence of a gene-by-environment interaction, in which an individual's response to environmental insults is moderated by his or her genetic make-up.

It is striking that recent research in epigenetics adds a new dimension to this knowledge, although the results of epigenetic studies are still controversial. "In summary, we conclude that the totality of the evidence on G x E is supportive of its reality but more work is needed to understand properly how 5-HTT allelic variations affect response to stressors and to maltreatment" (Rutter 2009, 1288).

(Leuzinger-Bohleber 2010a, b).[6] Another finding by Suomi (2010) is also highly relevant to us psychoanalysts. He was also able to show that undoing the separation trauma in baby monkeys might "undo" the neurobiological and behavioral damages again. This is, of course, a revolutionary finding for all forms of early prevention and psychotherapy, even though his results are based on studies involving animals. For psychoanalysts, these interdisciplinary findings are a strong motivating force behind our initiative in early prevention with at-risk families in several ongoing studies at the Sigmund-Freud-Institute (see www.sigmund-freud-institut.de).

3.4 Summary

What does clinical psychoanalysis have to offer neuroscientists? I decided to present my experiences concerning this question mainly through a clinical example from psychoanalysis with a severely depressed, traumatized patient instead of summarizing the wealth of psychoanalytical literature on trauma and depression, a task which is, of course, also an important contribution from the field of psychoanalysis to its neighboring disciplines (see Leuzinger-Bohleber et al. 2008a; Leuzinger-Bohleber in press a, b; Bohleber and Leuzinger-Bohleber in press). I tried to illustrate that remembering traumatic childhood experiences can only recur in a new interaction with a "meaningful other" (that is to say in the transference to the analyst). A situative, constructive understanding of interactions is the precondition for

---

[6] I agree with Goldberg (2009) when he states: "It is time that the dialogue of the deaf between psychiatric geneticists and psychotherapists came to an end: exciting progress has been made in understanding the interaction between our genetic constitution and social environment that either allow genes to manifest themselves in the phenotype, or suppress them altogether" (236). His conclusions, after having given an overview of the current state of research in this field, are highly relevant: "In humans, the effect of maternal care on hippocampal developments have so far been demonstrated (in females, but not in males). The effects of the environment in promoting gene expression appear to be supported by work showing that the extent of abnormalities in a particular gene, responsible for the metabolism of an important inhibitory neurotransmitter (serotonin), can be shown to be responsible for the sensitivity of the adult to external stress. This gene is also related to the likelihood of secure attachment. Thus the abnormalities observed in the rat also appear to apply to the human as well. Similarly, abnormalities in another gene, responsible for the neurotransmitter monoamine oxidase A, is associated with the sensitivity of the infant to the harmful effects of physical punishment—with gene normal, the relationship is fairly weak, but if it is abnormal, then anti-social behavior results." (244).
Goldberg concludes his overview of newer studies in these fields, "These interactions between gene and environment, between behavior and genotype are important in the way they provide explanations of how the many different features that make-up the 'depressive diathesis' arise. However, they have a much broader significance. They provide a possible pathway by which changing interpersonal and cultural factors across generations can be both the cause and the effect of a genotype, and through which changes in human culture might possibly be operating as an accelerator of evolutionary processes. In summary, we see that adverse environmental conditions are especially harmful to some particular genotypes, leaving the remainder of the population relatively resilient. Research in this area is expanding very fast—and we may expect many more advances in the years to come [...]" (244 f.).

remembering! Remembering is dependent on a dialogue in the inner and outside reality with an object, an interactive process, an integrative, 'embodied' experience between two persons. Ms. M. would not have been able to remember her traumata alone by herself lying in bed at home (see also results from recent trauma research, e.g., Brenneis 1994, 1997; Kihlstrom 1994; Van der Kolk et al. 1996; Leuzinger-Bohleber and Pfeifer 2002; Leuzinger-Bohleber et al. 2008b, 2010; Person and Klar 1994; Leuzinger-Bohleber 2010a, b, in press a, b). Thus, the unconscious but determining influence of early trauma can only be studied through intensive work with traumatized patients in the psychoanalytic situation and enables psychoanalysis to communicate these discoveries in an interdisciplinary dialogue with developmental and epigenetic researchers for example.

On the other hand, it seems that clinical research in psychoanalysis from the last few decades has gained interdisciplinary support from current biologically oriented memory research. And this psychoanalytic research postulates more and more radically that only by working through traumatic experiences in early object relations in the transference following a specific technique of treatment can we achieve a structural change in our patients. It seems that the findings from neuroscience and embodied cognitive science prove to be "externally coherent" (Strenger 1991) with psychoanalytic conclusions. In this respect, the interdisciplinary dialogue between psychoanalysis and neuroscience can be fruitful and innovative. I was not able to touch upon the ambitious epistemological and methodological problems that are, of course, connected to this dialogue. I have, however, discussed them in other papers (see e.g. Leuzinger-Bohleber in press a, b; Leuzinger-Bohleber et al. 2013). It is interesting that the old psychoanalytic dialectic that attempts to understand psychic phenomena in the complex interplay between mind and body, genetics and environment, biology and sociology, is again a central dimension in the contemporary dialogue between psychoanalysis and the neurosciences. Trauma and depression, as I tried to illustrate in the psychoanalysis with Ms. M., have unconscious, 'embodied' determinants as well as societal ones: Ms. M. was one of many traumatized children from the World War II period, whom German psychoanalysts have had in treatment. Understanding the complex and highly individual interactions between all these unconscious determinants in the psychoanalytical situation has proven to be indispensible for overcoming depression and for at least diminishing the influence of early trauma. In my experience the interdisciplinary dialogue with neuroscientists or embodied cognitive scientists can even aid in discovering and understanding the unconscious worlds of our psychoanalytic patients.

I would like to conclude with a personal remark. Based on my experiences a productive interdisciplinary dialogue also can be looked at from a psychoanalytical perspective. This can only happen if the participating researchers share a mature, high level of psychic functioning, and if they experience their interdisciplinary partner as an independent 'Other,' that is, as separate and as an object (and specifically not as a 'self-object') in a psychoanalytical sense. In this case, curiosity and a spirit of joint investigation can occur and overcome the desire or fear of being consumed or 'eaten up' by the partner and thereby losing one's own self and identity. Only in this case can the "Eros" of interdisciplinary cooperation defeat its counterpart "Thanatos," as Freud taught us in a different context.

References

Argelander, H. (1987). *Das Erstinterview in der Psychotherapie* (3rd ed.). Darmstadt: Wissenschaftliche Buchgesellschaft.
Benecke, C., & Krause, R. (2001). Fühlen und Affektausdruck: Das affektive Geschehen in der Behandlung von Herrn P. *Psychotherapie und Sozialwissenschaft, 3*, 52–73.
Bifulco, A. T., Brown, G. W., & Harris, T. O. (1987). Childhood loss of parent, lack of adequate parental care and adult depression: A replication. *Journal of Affective Disorders, 12*, 115–128.
Bohleber, W. (2000). Die Entwicklung der Traumatheorie in der Psychoanalyse. *Psyche – Z Psychoanal, 54*, 797–839.
Bohleber, W. (2005). Zur Psychoanalyse der Depression. *Psyche – Z Psychoanal, 59*, 781–788.
Bohleber, W. & Leuzinger-Bohleber, M. (in press). The special problem of interpretation in the treatment of traumatized patients. Will be published in *Psychoanalytic Inquiry*, 2014.
Bollas, C. (1992). *Being a character: Psychoanalysis and self experience*. New York: Hill and Wang.
Brenneis, C. B. (1994). Belief and suggestion in the recovery of memories of childhood sexual abuse. *Journal of the American Psychoanalytic Association, 42*, 1027–1053.
Brenneis, C. B. (1997). *Recovered memories of trauma: Transferring the present to the past*. New York: International Universities Press.
Caspi, A., Sugden, K., Moffitt, T. E., Taylor, A., Craig, I. W., Harrington, H. L., McClay, J., Mill, J., Martin, J., Braithwaite, A., & Poulton, R. (2003). Influence of life stress on depression: Moderation by a polymorphism in the 5-HTT gene. *Science, 301*, 386–389.
Clancey, W. J. (1993). The biology of consciousness: Comparative review of Israel Rosenfield, The strange, familiar and forgotten: An anatomy of consciousness and Gerald M. Edelman, Bright Air, Brilliant Fire: On the matter of the mind. *Artificial Intelligence, 60*, 313–356.
Fergusson, D. M., & Mullen, P. E. (1999). *Childhood sexual abuse: An evidenced-based perspective*. London: Sage.
Fischer, G., & Riedesser, P. (1998). *Lehrbuch der Psychotraumatologie*. München/Basel: Reinhardt.
Fonagy, P., & Target, M. (1997). Perspectives on the recovered memories debate. In J. Sandler & P. Fonagy (Eds.), *Recovered memories of abuse: True or false?* (pp. 183–217). London: Karnac Books.
Fuchs, T., Sattel, H. C., & Henningsen, P. (Eds.). (2010). *The embodied self: Dimensions, coherence and disorders*. Stuttgart: Schattauer.
Goldberg, D. (2009). The interplay between biological and psychological factors in determining vulnerability to mental disorders. *Psychoanalytic Psychotherapy, 23*, 236–247.
Green, A. (1999). The greening of psychoanalysis: André Green in dialogues with Gregorio Kohon. In G. Kohon (Ed.), *The dead mother: The work of André Green* (pp. 10–58). London/New York: Routledge.
Hill, J. (2009). Developmental perspectives on adult depression. *Psychoanalytic Psychotherapy, 23*, 200–212.
Hill, J., Pickles, A., Burnside, E., Byatt, M., Rollinson, L., Davis, R., & Harvey, K. (2001). Sexual abuse, poor parental care and adult depression: Evidence for different mechanisms. *British Journal of Psychiatry, 179*, 104–109.
Hinshelwood, R. D. (1991). *A dictionary of Kleinian thought* (2nd ed.). London: Free Association Books.
Kendler, K. S., Gatz, M., Gardner, C. O., & Pederson, N. L. (2006). A Swedish national twin study of lifetime major depression. *American Journal of Psychiatry, 163*, 109–114.

Kihlstrom, J. F. (1994). Hypnosis, delayed recall and the principles of memory. *International Journal of Clinical and Experimental Hypnosis, 42*, 337–345.
Laub, D. (2005). Traumatic shutdown of narrative and symbolization: A death instinct derivative? *Contemporary Psychoanalysis, 41*, 307–326.
Leuzinger-Bohleber, M. (2010a, October 23). *Depression and trauma – a transgenerational psychoanalytical perspective*. Unpublished paper, 1st Asian conference of the IPA: Freud and Asia. evolution and change: Psychoanalysis in the Asian context. Peking.
Leuzinger-Bohleber, M. (2010b, June 3–6). *Transgenerational trauma – an unexpected clinical observation in extra-clinical studies*. Lecture at the 36th annual meeting of the Canadian psychoanalytic society: 100 years on the couch – Psychoanalysts worldwide celebrate centenary of International Psychoanalytical Association. Toronto.
Leuzinger-Bohleber, M. (in press a). Working with severely traumatized, chronic depressed patients. Will be published in the *International Journal of Psychoanalysis*, 2014.
Leuzinger-Bohleber, M. (in press b). *Finding the body in the mind. Freud's scientific dreams: Psychoanalysis-neurosciences-embodied cognitive science in dialogue. Inspiration for clinical practice, theory and research*. Will be published by Karnac Books, London, 2015 (IPA publications series).
Leuzinger-Bohleber, M., & Pfeifer, R. (2002). Remembering a depressive primary object: Memory in the dialogue between psychoanalysis and cognitive science. *International Journal of Psychoanalysis, 83*, 3–33.
Leuzinger-Bohleber M. & Pfeifer, R. (2011, June 25). Minding the traumatized body – Clinical lessons from embodied intelligence. 12th international neuropsychoanalysis congress "Neuropsychoanalysis: Minding the body". Berlin.
Leuzinger-Bohleber, M., & Pfeifer, R. (2013). Embodiment: Den Körper in der Seele entdecken – Ein altes Problem und ein revolutionaires Konzept. In M. Leuzinger-Bohleber, R. N. Emde, & R. Pfeifer (Eds.), *Embodiment: Ein innovatives Konzept für Entwicklungsforschung und Psychoanalyse*. Göttingen: Vandenhoeck & Ruprecht.
Leuzinger-Bohleber, M., Rüger, B., Stuhr, U., & Beutel, M. (2002). *"Forschen und Heilen" in der Psychoanalyse. Ergebnisse und Berichte aus Forschung und Praxis*. Stuttgart: Kohlhammer.
Leuzinger-Bohleber, M., Roth, G., & Buchheim, A. (Eds.). (2008a). *Psychoanalyse – Neurobiologie – Trauma*. Stuttgart: Schattauer.
Leuzinger-Bohleber, M., Henningsen, P., & Pfeifer, R. (2008b). Die psychoanalytische Konzeptforschung zum Trauma und die Gedächtnisforschung der Embodied Cognitive Science. In M. Leuzinger-Bohleber, G. Roth, & A. Buchheim (Eds.), *Psychoanalyse, Neurobiologie, Trauma* (pp. 157–171). Stuttgart: Schattauer.
Leuzinger-Bohleber, M., Röckerath, K., & Strauss, L. V. (Eds.). (2010a). *Depression und Neuroplastizität. Psychoanalytische Klinik und Forschung*. Frankfurt: Brandes & Apsel.
Leuzinger-Bohleber, M., Bahrke, U., Beutel, M., Deserno, H., Edinger, J., Fiedler, G., Haselbacher, A., Hautzinger, M., Kallenbach, L., Keller, W., Negele, A., Pfenning-Meerkötter, N., Prestele, H., Strecker von Kannen, T., Stuhr, U., & Will, A. (2010b). Psychoanalytische und kognitiv-verhaltenstherapeutische Langzeittherapien bei chronischer Depression: Die LAC-Depressionsstudie. *Psyche – Z Psychoanal, 64*, 782–832.
Leuzinger-Bohleber, M., Emde, R. N., & Pfeifer, R. (Eds.). (2013). Embodiment – ein innovatives Konzept für Entwicklungsforschung und Psychoanalyse. In *Schriften des Sigmund-Freud-Instituts. Reihe 2: Psychoanalyse im interdisziplinären Dialog* (Vol. 17). Göttingen: Vandenhoeck & Ruprecht.
Person, E. S., & Klar, H. (1994). Establishing trauma: The difficulty distinguishing between memories and fantasies. *Journal of the American Psychoanalytic Association, 42*, 1055–1081.
Pfeifer, R., & Leuzinger-Bohleber, M. (1986). Applications of cognitive science methods to psychoanalysis. A case study and some theory. *International Review of Psycho-Analysis, 13*, 221–240.

Pfeifer, R., & Leuzinger-Bohleber, M. (1989). Motivations- und Emotionsstörungen: ein Cognitive Science Ansatz. Teil I: Theoretische Überlegungen. *Zeitschrift für Klinische Psychologie, Psychopathologie und Psychotherapie, 37*, 40–73.

Pfeifer, R., & Leuzinger-Bohleber, M. (1992). A dynamic view of emotion with an application to the classification of emotional disorders. In M. Leuzinger-Bohleber, H. Schneider, & R. Pfeifer (Eds.), *"Two butterflies on my head ..." Psychoanalysis in the interdisciplinary scientific dialogue* (pp. 215–245). New York: Springer.

Robertson, J., & Robertson, J. (1975). Reaktionen kleiner Kinder auf kurzfristige Trennung von der Mutter im Lichte neuer Beobachtungen. *Psyche* Heft 7, 29. Jahrgang, 626–665.

Rutter, M. (2009). Gene-environment interactions: Biologically valid pathway or artifact? *Archives of General Psychiatry, 66*, 1287–1289.

Sandler, J., & Sandler, A.-M. (1997). A psychoanalytic theory of repression and the unconscious. In J. Sandler & P. Fonagy (Eds.), *Recovered memories of abuse: True or false?* (pp. 173–181). London: Karnac Books.

Stern, D. N. (1995). *Die Mutterschaftskonstellation. Eine vergleichende Darstellung verschiedener Formen der Mutter-Kind-Psychotherapie.* Stuttgart: Klett-Cotta.

Strenger, C. (1991). *Between hermeneutics and science: An essay on the epistemology of psychoanalysis.* New York: International Universities Press.

Suomi, S. (2010, February 7). *Trauma and epigenetics.* Unpublished paper given at the 11th Joseph Sandler research conference: Persisting shadows of early and later trauma. Frankfurt am Main.

Van der Kolk, B., McFarlane, A. C., & Weisaeth, L. (Eds.). (1996). *Traumatic stress: The effects of overwhelming experience on mind, body, and society.* New York: Guilford Press.

Chapter 4
Embodiment in Simulation Theory and Cultural Science, with Remarks on the Coding-Problem of Neuroscience

Sigrid Weigel

Abstract The chapter discusses the role of cultural sciences (*Kulturwissenschaften*) in bridging the still existing gap between neuroscience and psychoanalysis against the background of the rising interest in *mirror mechanisms, mimetic functions, imitation, simulation, embodiment, language and metaphors* as well as their proximity to figures that have long belonged to the traditional subjects of the humanities. Knowledge on human expressions within a cultural-historical timeframe may be adduced to bridge the gap between data from empirical research and the evolution of the brain based on speculative knowledge. Weigel concentrates on certain unsolved epistemological problems to be reconsidered in light of knowledge from cultural science. Drawing on Freud and Walter Benjamin, she examines the quantity-quality-problem and the coding problem in simulation theory with reference to theories of language, gesture and image in the humanities. An example completes the chapter's presentation of the cultural production of 'compassion' in medieval times in relation to the concept of empathy in recent neuroscience.

Keywords *Kulturwissenschaften* • Simulation theory • Mirror neurons • Mimesis • Coding • Language • Cultural history • Empathy • Compassion

4.1 Prologue

This article begins with a trouvaille from the *Denktagebuch*, a sort of intellectual notebook, of Hannah Arendt, the famous German-Jewish philosopher (1906–1975). Arendt's publications include a most profound book on the *Human Condition* (1958, in German *Vita Activa*, 1960) in which she develops the idea of *acting/ Handlung* as the crucial realm of intersubjectivity and humanity. This realm is based in the space between human beings, a literal *interest* of togetherness. It is only in

---

S. Weigel (✉)
Zentrum für Literatur- und Kulturforschung, Schützenstrasse 18, D-10117 Berlin, Germany
e-mail: direktion@zfl-berlin.org

© Springer International Publishing Switzerland 2016
S. Weigel, G. Scharbert (eds.), *A Neuro-Psychoanalytical Dialogue for Bridging Freud and the Neurosciences*, DOI 10.1007/978-3-319-17605-5_4

this space, only in the relationship to others, that the full sense of the self, including a person's involuntary expressions, manifests itself. It is the same realm in which the moral, social and political life is created.

The 44-year-old Arendt's notebook contains the following entry:

> Nothing reveals more of the ambiguity of language than the metaphor. I, for example, have been using the metaphor 'I open my heart' my whole life, without ever having felt the actually physical sensation. Only once I've experienced this physical sensation, I realised how often I had lied in the past. Yet, how could I have experienced the fundamental truth of this physical sensation, had not language prepared me with its metaphor for the significance of the event? (22. December 1950)[1]

The entry discusses the mutual transferal between emotions and bodily sensations by reflecting on the role of language as a mediator that embodies the psyche and gives psychical meaning to physiological feelings. Since the phrase 'open heart' belongs to a register of long-established metaphors, these reflections concern the comprehension of body-metaphors and their role in the *shared meaningful space of experiences* (Gallese 2009a, 527), i.e., language as a transmitter of experiences and memory in cultural history.

4.2 Introduction: Point of Departure and Field of Intervention

The purpose of what follows is an intervention from a cultural-historical perspective in the dialogue of psychoanalysis and the neurosciences. To start, I first need to present some background about the field from which my intervention comes.

4.2.1 On the Neurosciences

The point of departure here is the rising interest in *mirror mechanisms, mimetic functions, imitation, simulation, embodiment, language and metaphors*—all of which emerged after, and as a result of, the discovery of the *MNS (Mirror Neuron System)*. For a scholar coming from the fields of philology and *Kulturwissenschaften*, best translated perhaps as cultural sciences, this entire register of concepts sounds quite familiar. Research in the history of culture has always been based in the conviction of a *close connection between mind, imitation, gestures and acting.*

---

[1] In nichts offenbart sich die eigentümliche Vieldeutigkeit der Sprache […] deutlicher als in der Metapher. So habe ich zum Beispiel ein Leben lang die Metapher 'es öffnet sich mir das Herz' benutzt, ohne je die dazu gehörende physische Sensation erfahren zu haben. Erst seit ich die physische Sensation kenne, weiss ich, wie oft ich gelogen habe, […] Wie aber hätte ich je die Wahrheit der physischen Sensation erfahren, wenn die Sprache mit ihrer Metapher mir nicht bereits eine Ahnung von der Bedeutsamkeit des Vorgangs gegeben hätte? (Arendt 2002, 46).

Therefore, this new approach in the neurosciences, as brought about by the discovery of the mirror neurons and the subsequent emphasis on simulation theory, enactive mind, human semantics, gestural language, symbolic thought and metaphors (Fonagy and Target 2007), opens a fascinating possibility of a dialogue between the fields. Through the discussion of figures like embodied simulation, shared interpersonal space, intercorporeity and empathy (Gallese 2009b), the epistemology of neuroscience has overcome its long-standing *underestimation of imitation*. This status was underpinned by a notion of the relation between brain activity and motoric action that was too mechanistic with the computer as the iconic metaphor for the brain and the body but an appendage of brain functions.

Philology and cultural history are long-established fields of research occupied with the detailed analysis of textual, oral, gestural, visual and other expressions with well thought-out theories of mimesis, imitation, imagination, perception and deciphering. Therefore, it is time for this field of knowledge to take part in the effort of bridging neurosciences and psychoanalysis.

In order to give an impression of this field's relevance, one should remember the German concept of *Geisteswissenschaft* for the humanities that literally means 'mind-science'. One could also think about one of the central ideas of literary theory, namely imagination (in German *Einbildung*), a concept that includes the word 'image' (*Bild*) and literally means an incorporation of an image (of an outer event, object, piece of art) into the soul and the mind, in other words: embodied perception. To get an idea of the history of knowledge on imitation and embodied simulation, one might initially think of three *classical points of reference* from the cultural sciences:

(1) Aristotle's idea of man as *zôon mimêtikôtaton;*
(2) the eighteenth-century concept of *Bildung*, i.e., the idea that a person's development is triggered by the inscription of images (an ontogenetic concept of epigenesis), as when the shape of an antique sculpture gets embodied and inscribed into the mind or soul as a *signature* or trace, as Karl Philip Moritz explains in his *Signature of Beauty* and in his reflections on the "bildende" mimesis of beauty (Moritz 1968)[2];
(3) the concept of *pathos formula*, sometimes also called *dynamogram*, developed around 1900 by the German-Jewish art historian Aby Warburg, the founder of the K.B.W. (Kulturwissenschaftliche Bibliothek Warburg), described as a kinetic image or image of bodily movements, and interpreted as a mediator for the expression of emotion between people, spaces and time periods (Warburg 2010);
(4) the concept of *Einfühlung* developed by authors of aesthetic theory in the second half of the nineteenth century in Germany, which addresses the issue of animation (see Papapetros 2012) and questions as to why and how human beings transpose their emotions to pieces of art, artefacts, other inorganic things,

---

[2] For a more detailed analysis of this role of images see the second chapter on traces and lines in Weigel 2015.

and even to forms of architecture. With its translation into the English-American psychology, this concept has provided a term for ideas on *empathy*, which is currently at the center of research in psychology, socio-neuroscience, and anthropology (Lux and Weigel (forthcoming) 2015).

4.2.2 On Psychoanalysis

The field of philology can look back at its long, stable and fruitful relationship to psychoanalysis because the writings of Sigmund Freud, albeit without their clinical and therapeutic dimensions, belong to the humanities' and cultural sciences' body of knowledge. Thus, Freud's work can be read as a theory of subjective and cultural meaning that provides figures, explanations and laws for their genesis. During the extended period in which psychoanalysis has been without an academic home in universities (at least in German universities and in many other countries) it survived as a theory partly because many of Freud's texts were integrated into the curricula of several departments of literature and cultural science.

Since the *cultural sciences* (*Kulturwissenschaft*) approach is unfamiliar in English I have to explain this field to avoid it being mistaken for cultural studies. Whereas the latter is mainly occupied with questions of class, race and gender and with interpretations of popular or mass media culture, *Kulturwissenschaft* is concerned with conceptual, visual and epistemological issues in the history of knowledge and culture. It goes back to an intellectual undertaking around 1900 dedicated to developing an approach beyond the divide of the 'two cultures' of science and the humanities. Concerned with the *cultural production of meaning*, this theoretical venture emerged from the (for the most part disillusioned) attempt to deal with empirical methods and approaches that borrowed from the natural sciences to analyze cultural phenomena.

An interesting example comes from the Physiological Institute Franz Brücke in Vienna, where Sigmund Exner (1846–1926) attempted to measure the probable weight of flying objects (like angels and other heavenly figures) in paintings (Figs. 4.1 and 4.2). Once he realized that he would not be able to reach any reasonable results this way, he decided to change the approach and instead to deal with memory images (Weigel 2007).

The venture of *Kulturwissenschaft* was, at the same time, based on a critique of the metaphysical concept of 'Geist' in the establishment of *Geisteswissenschaft* (by Wilhelm Dilthey). Scholars in the cultural sciences were instead occupied with analyzing material culture, i.e., practical, symbolic and corporeal modes of expressing emotions and meaning, and thus interested in embodied and enacted modes of all sorts of human articulations. Aby Warburg, Sigmund Freud, Georg Simmel, Walter Benjamin and Helmuth Plessner belong to this project, to name just a few authors. This means that Sigmund Freud's works constitute both a fundamental theoretical source for *Kulturwissenschaft* as well as an object and body of ideas to be studied in terms of their epistemo-historical and cultural premises.

4 Embodiment in Simulation Theory and Cultural Science, with Remarks…

Fig. 4.1 Sigmund Exner (1882). *Die Physiologie des Fliegens und Schwebens in den Künsten*

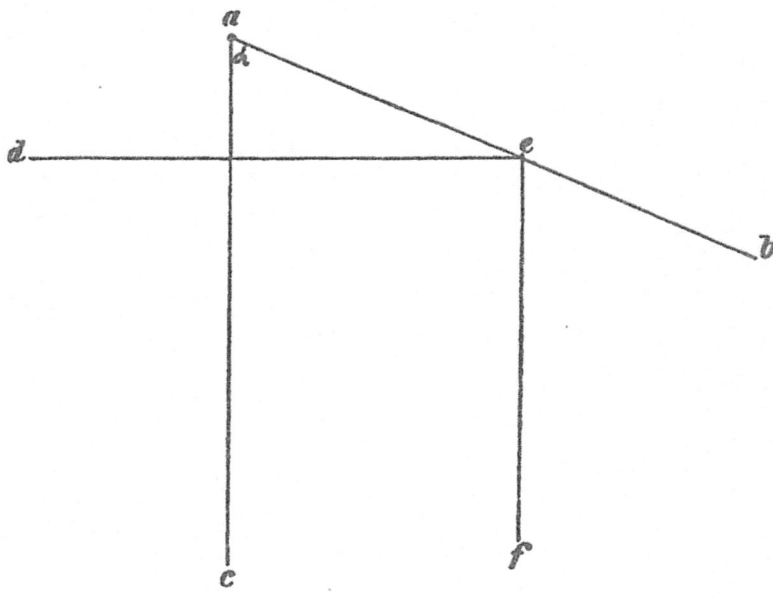

Fig. 4.2 Diagram by Exner to the iconic picture in Fig. 4.1.

4.3 Epistemological Problems Part I: The Quantity-Quality-Gap

The attempts by the above-mentioned authors from around 1900 are, in my eyes, of acute contemporary interest because they correspond with the current configuration of neuro-psychoanalysis, an ambitious project in which both fields are challenged to find ways of bridging their radically different epistemologies of, on the one hand, an empirically based science and, on the other, an approach occupied with language and individual memory.

Today, the epistemological situation in neuroscience is structured by empirical research, which is based on the use of advanced *technology* (brain imaging, etc.) and highly differentiated *experimental studies* with fascinating empirical findings, and by the theoretical debate surrounding the *construction* of models (of the self, the brain-body-relation, etc.). The problem with visualizations of the brain can be observed in multiple scientific presentations, in which the images actually level the differences between empirical findings, iconic models, and assumptions about inaccessible physiological regions. Since the visual representations do not distinguish between data and models, between neurological "facts" (i.e., the visualization of measured neurological activities) and graphic representations of theoretical assumptions, many *images in neuroscientific papers* or *lectures* are symptoms of encapsulated, unsolved problems (Weigel 2004; third chapter in Weigel 2015).

Although the technological tools providing the possibility to literally "look into the brain" have increased enormously in number since Freud's time, fundamental epistemological problems (still) exist in current neuroscience, problems that are comparable to the dual trait Sigmund Freud had to deal with when he developed his metapsychology. The recent model of the self—for example, the distinction between proto-self, core self and autobiographical self (Damasio 2010)—has, in epistemological terms, the same constructed character as Freud's models of the psychic apparatus. However, in contrast to Freud's constant methodological reflections and recurrent discussion of the inevitability of speculation and assumption, the constructed character of today's neuro scientific models, a sort of *metamindology*, gets lost for the most part in descriptions of the brain and its anatomical and neuronal parts and functions.

Still, one of the gravest problems in the neurosciences today is not far removed from the theoretical problems Freud so lucidly discussed in his *Project for a Scientific Psychology/Entwurf einer Psychologie* (1895), namely the gap between *quantitative-empirical methods* and the *question of quality* or rather *meaning*. In this unfinished and posthumously published manuscript, he reflects upon the incompatibility between the "quantitative problem" or "biological position" (neurons as material/physiological bearer of psychic processes) and the "problem of quality" (meaning). In the introduction to the *Project* he expresses his intention to present a psychology based on the natural sciences (*eine naturwissenschaftliche Psychologie*):

> The intention is to furnish a psychology that shall be a natural science: that is, to represent psychical processes as quantitatively determinate states of specifiable material particles, thus making those processes perspicuous and free from contradiction. Two principal ideas are involved:

1. What distinguishes activity from rest is to be regarded as Q, subject to the general laws of motion.
2. The neurons are to be taken as the material particles. (S.E. I, 295)

This methodological assumption strikingly resembles the *current neuroscientific approach*, namely in the difference between repose and activity as the fundamental entity used in all measurements for brain scanning and mind mapping methods.

In his chapter on the "problem of quality" Freud reflects that the inquiry into *content* confronts us with totally different problems. He argues that what we know about the content or meaning of the quantities we only know from our consciousness. Consciousness gives us

> what are called *qualities* – sensations which are *different* in a great multiplicity of ways and whose *difference* is distinguished according to its relations with the external world. Within this difference there are series, similarities and so on, but there are in fact no quantities in it. It may be asked *how* qualities originate and *where* qualities originate. (Ibid., 308)

In the remarks that follow this passage, Freud claims that these are questions that need a thorough examination, a claim that is still valid today.

In his later works after the invention of psychoanalysis when he was mainly dealing with corporeal and linguistic symptoms of meaning, *the economic principle* within the metapsychological triad (dynamic, topic, economic)—namely, his references to a higher or lower level of energy, pressure, or drive and to the *intensity* or magnitude of arousal or excitation—function as a proxy for quantity within a qualitative approach. Whereas Freud's research went from neurology to psychoanalysis, the current constellation is organized the other way round, since the neuroscientific discourse has returned to Freud and psychoanalysis during the last decades. Therefore, the crucial epistemological problem might also be identified in reverse: *What are the proxies for qualities in a dominantly quantitative approach?*

In contemporary research these are mainly correlated indicators, which are examined in experimental activities through observation and the reports of probands. But in order to combine these correlative indicators with the findings of the brain scan one has to isolate them from individual, social and cultural contexts, transfer them into algorithms and define or fix their meaning. In order to assimilate these qualities into quantitative methods as much as possible, one has to reduce the complexity of meaning by isolating single indicators and measurable entities. The definition of entities that function as proxies for quality or meaning has recently tended to concentrate on two terms: *map* and *code*.

Whereas *mapping* is a spatial figure of knowledge that relies on anatomical evidence in order to identify patterns of neurological activities (e.g., Bud Craig's "global emotional movement" defined as a map of energy representation, Craig 2011), the *code* stems from a linguistic or semiotic knowledge base and was introduced into the life sciences decades ago. The metaphorical state of the term has now mostly been forgotten (see the chapter on the genetic code in Weigel 2006). In neuroscience the term code is presently used as a mediator between neurological facts and the meaning of feelings and memory. Due to the increasing relevance of MNS, simulation theory and embodied intersubjectivity, language and meaning have entered the core of neuroscientific research. As a result the crucial epistemological problem has shifted from a quantity-quality gap to a *coding problem*.

4.4 Epistemological Problems Part II: The Coding Term in Simulation Theory

When discussing language, metaphors and the semantics involved in the mirror function of embodied simulation, the current neuroscientific discourse regularly refers to the concept of code and coding. I will mention just a few examples:

The idea of a "vocabulary of acting" in Giacomo Rizzolatti and Corrado Sinigaglia's mirror neuron book *Mirrors in the Brain. How Our Minds Share Actions and Emotions* (2008) constructs the movements involved in the MNS as a sort of lexicon, i.e., a body of words with conventional meanings. Vittorio Gallese conceives the MNS as a sort of coding system when, for example, he speaks of "brain areas that preferentially code for esthetic stimuli" (Gallese and Di Dio 2011, 6) or when talking about how mirror neurons code the meaning of others' actions (Gallese 2009a, 521). Peter Fonagy and Mary Target, in contrast, discuss a "dual coding of language" and distinguish the "dictionary meaning of a word" from a second meaning; but they also conceive the latter as a code on two different levels: the subjective and the evolutionary. On the subjective level, that is as a "human semantics that map a person's cumulative experience", it is defined as "*sense*, as opposed to *meaning*, that is embodied and encoded through experiences of the physical body" (Fonagy and Target 2007, 433; ref. Klin & Jones 2007). On the evolutionary level, it is seen as a "second, embodied, physical-experience-based coding system built into language by its evolutionary history" that goes back to gestures (Fonagy and Target 2007, 433; ref. I. Fónagy 2000). In this way the MNS, simulation theory and its Interpersonal Interpretative Function (IIF) are all conceptualized as an *encoding-decoding system*.

The reference to the concept of *code* has also been used to refer to a 'second level of meaning', i.e., language that exists beyond the conventional system of words as signs. This other linguistic realm lies beyond the arbitrary language of communication, which obscures the fact that corporeal expressions, gestures and other articulations do not present fixed meanings. Although they have partly been shaped by history and culture and learned through experience, acculturation, and habituation, the decodable part of them is very limited. This applies not just to gestures, expressions of emotions, facial expressions and the like, but also to certain phenomena within verbal language, especially phenomena linked to the human voice: intonation, rhythm, pitch, breaks, breathing, stuttering, etc. Julia Kristeva has described this field of uncodable meaning as a realm of the *semiotic*,[3] in contrast to the *symbolic*, where meaning is determined by the laws of language (Kristeva 1984). It is no coincidence that this register of expression has a great deal in common with Freud's language of the unconscious. Indeed, this concept marks the intersection where research in literary theory and cultural science meets psychoanalysis. Also the focus on antenatal and neonatal life in developmental psychology has spurred research on

---

[3] Not to be confused with the more established term of semiotics as a theory of signs defined by Charles Sanders Peirce (1839–1914).

these elements of language, especially in connection with the mother's voice and its impact on the embryo and the new-born child (Mancia 2006, 24, ref. to Kolata 1984; Mehler & Bertoncini 1978).

Contrary to this attention on complex expressions of meaning, the coding paradigm within the neuroscientific research community has so far mainly been informed by *linguistics*. Within the broad field of linguistics the coding model refers either to speech act theory, to communication theory informed by information technology, or to formal linguistics underpinned by mathematics and cognitive or analytical philosophy. The understanding of language in cultural science is different; it is related more to continental philosophy. At this point an excursus on language theory cannot be avoided.

It is rare to come across linguistic theories open to expressions beyond verbal language, signs and code. One rare exception can be found in the work of Ivan Fónagy and the approach to gestures in his theory of language, to which Fonagy and Target 2007 refer. *Linguistics*, regardless which camp of the "linguistic war" (Chomsky—Lakoff) you are dealing with, conceives of language as a *system of signs and rules* that generates sentences and enunciations. It mainly refers to spoken language and conceptualizes other modes of expression that are analogous to verbal language. An *encoding-decoding system*, in the strict sense of the words, comes from *information theory* (or communication theory); it is construed as a one-to-one model in which language is seen as an arbitrary, conventional system. The coding of the sender and the decoding by the receiver are constructed as complementary actions whereby differences that occur between the encoded and decoded message can only be described in terms of *disturbance* or failure.

While studying some recent papers on simulation theory, I found one idea especially problematic: George Lakoff's idea of a "conceptual metaphor". One could easily argue that because of its basis in cognitive theory his concept of metaphor cannot provide a helpful means for addressing neuro-psychoanalytical questions developed in the wake of the 'emotional turn' in neuroscience. However, it is more complicated than that. The idea of *metaphor* (from the Greek μεταφορά, to transfer; literally, to carry from one point to another) can only emerge when it becomes distinct from direct designation or denotation, i.e., when it is distinguished from a *conceptual term* (*Begriff*). Seen from the longue-durée perspective of cultural history, one has to add that the division between metaphors and concepts depends on the prior existence of an imagistic consciousness, a thinking that occurs before and below the separation of direct and metaphorical expression. Thinking in images is the beginning and foundation of language.

The metaphor belongs to the traditional key concepts of humanities. Consequently, there are numerous works on metaphors that provide detailed analyses particularly on the distinctions between different kinds and fields of metaphorical thought. In his research on the history and theory of metaphors Hans Blumenberg (1920–1996), for example, starting with his *Paradigms for a Metaphorology* (2010), broadly examined this realm of what he calls *unconceptual thinking*, a common and often creative kind of thinking without strict definable concepts, ideas or terms. His work is dedicated to differentiating between various fields, qualities and functions of metaphors. He analyzed the field of *Daseinsmetaphern* (metaphors of being), i.e., figures that structure a whole realm of experience; his most famous example is the shipwreck (Blumenberg 1996). Although this field might be compared to Lakoff's 'conceptual metaphors', in light of Blumenberg's theory such linguistic images are not to be seen as universal; instead their generation depends on experiences connected to the environment of different cultures (land, sea, agriculture, industrial, rural areas, cities, etc.). The field of *absolute metaphors* is quite different. Here metaphors cannot be translated into concepts at all. These metaphors are often used in science and are particularly useful for fields not yet fully conceived or for fields of knowledge that concern interior, invisible or in other ways inaccessible phenomena. Needless to say, the mind and psyche belong to this field (Blumenberg 2007). Blumenberg was also one of the first scholars from the humanities to present an enlightening analysis of the metaphorical generation of the 'genetic code' (Blumenberg 1996). Set against his metaphorology many of Freud's terms may be described as absolute metaphors.

In comparison to Blumenberg's theory Lakoff's 'conceptual metaphor' is constructed as a more or less universal phenomenon. He considers metaphors as being an "unconscious symbolic thought" and as a "*natural mechanism* for relating concrete images to abstract meanings" (Lakoff 1997; italics added). Although his examples are taken from modern, everyday life and ordinary language, the construction of such a realm of quasi-natural metaphors has a lot in common with C. G. Jung's idea of the archetype. Since Lakoff discusses the conceptual metaphor in relation to his model of a 'cognitive unconscious', this idea seems at first glance to be of interest to the current dialogue between neuroscience and psychoanalysis. However, what he calls 'unconscious', when discussing the function of metaphors in "the normal mind" (Lakoff 1997, 90) and defining it as a sort of "fast, automatic, effortless—and completely normal" thinking, is quite different from the unconscious in psychoanalysis.

Especially with regard to the 'return to Freud' in neuroscience, one has to take care to work with terms that are as precise as possible. What phenomena have not been called *unconscious* in recent scholarship? But terminology matters! Since Freud's terminology of *conscious, unconscious,* and *preconscious* provides the central terminology of psychoanalysis, it makes no sense to use the same terms indiscriminately for other phenomena. We need clear, distinct and differentiated terms for what is *not conscious*, for example, whether a state is seen as *preconscious, unconscious, a-conscious, involuntary, automatic, reflexive*, etc. The small prefixes that supplement the conscious in this field of terminology produce an all or nothing differential, as George-Arthur Goldschmidt has shown in his readings of English and French translations of Freud's writings (Goldschmidt 2000).

The mode of metaphorical expression Lakoff describes as fast, automatic and effortless has been analyzed in the cultural sciences as a broad field of *involuntary* expression and acting. It concerns conventional meanings that are acquired not through an organized procedure of learning but instead through a sort of *habituation*—or to put it in the terms of simulation theory, as a visuo-acoustic-motoric embodiment. This takes place as a kind of involuntary acculturation, and the particular meanings depend on the specific historical and cultural environment that establishes the source domain of the metaphorical operation. The whole field of involuntary thought and action undergoes a great change due to developments in the technology, instruments and machines that invade everyday life. As Walter Benjamin discusses in his history of European modernity (*The Arcades Project*), the field of involuntary motor movements was massively extended during the nineteenth century when life became equipped with more and more instruments (Benjamin 2002).

But instead of criticizing the language theory of linguistics any further, I would like to suggest another approach to *language (in the broader sense)* that includes all sorts of textual, oral, visual, gestural, bodily and other expressions of affects and meaning.

4.5 Mimetic and Arbitrary Aspects of Language—The Time of Cultural History

A crucial question in simulation theory asks about the connection between visual and motor functions and its fundamental importance when it comes to perceiving the actions of others and the resulting motor activities in the perceiving person.

In his article on *Mirror Neurons, Embodied Simulation and the Neural Basis of Social Identification*, Vittorio Gallese poses the vitally important question as to "how the MNS develops in the course of development" (Gallese 2009a, 529). In order to find an answer, he discusses findings from experimental research on kinematic patterns during prenatal development in neonates and infants. This research deals with the *ontogenetic dimension* that is the only developmental dimension accessible to empirical research. In this context, he presents a study by Sotaro Shimada and Kazuo Hiraki (2006) that demonstrates "an action execution/observation matching system in 6-month-old human infants". Gallese reports on the findings as follows:

> Interestingly, this study showed that the sensory-motor cortex of infants (but not that of adult participants) was also activated during the observation of a moving object when presented on a TV screen. These findings suggest that during the early developmental stages, even non biological moving objects are 'anthropomorphized' by means of their mapping onto motor representations pertinent to the observers' acquired motor skills. (Gallese 2009a, 530)

And he summarizes these findings with the hypothesis that an innate rudimentary MNS, which is already present at birth, can be flexibly modulated by motor experience and gradually enriched by *visuomotor learning*.

What immediately attracted my attention when reading this passage was the anthropomorphism of objects in the infants' perception. From a cultural science perspective, this can be interpreted as an early stage of simulation based in a pre-existing mimetic culture. It reminds me of a beautiful piece in Walter Benjamin's autobiographic book *Berlin Childhood around 1900* (written in 1933) that does not consist of a developmental narrative but rather of a series of short scenarios or thought-images. In a piece entitled *Butterfly Hunt*, he describes a young boy's movements when chasing butterflies: his mimetic movements follow the movements of the fluttering creature as well as the echoes of an inner flittering elicited by his bodily movements: "the more I strove to conform, in all the fibers of my being, to the animal—the more butterfly-like I became in my heart and soul". (Benjamin

2002, 351) There are other passages in this book in which he describes a similar attitude towards non-living environments. He thus posits a fundamental mimetic faculty in infants condensed in the statement, "I was distorted by similitude with everything which was surrounding me." (The Mummerehlen, Benjamin 2002, 374) He also shows how this attitude gets constrained, subjected and overlapped by learning conventions and codes, both of language and acting, as well as by learning to distinguish between living beings and an organic world.

This refraining from mimetic action anticipates the findings of another empirical study, cited by Gallese: "Lepage/Théoret (2007) recently proposed that the development of the MNS can be conceptualized as a process whereby the child learns to refrain from acting out the automatic mapping mechanism linking action perception and execution." (Gallese 2009a, 530) Whereas psychoanalysis describes social learning as the inhibition of desires and drives, one may add that, as regards language, this ontogenetic process is accompanied by an inhibition of mimetic modes of expression within social and symbolic language. The innovative turn towards a *cultural historical approach* to this topic makes it possible to introduce *a third temporal dimension* that provides a supplement to the two perspectives of phylogenetic/evolutionary development, on the one hand, and ontogenesis, on the other.

Neuro-psychoanalysis is based on the idea of correspondences between ontogenesis and phylogenesis. *Ontogenetic development* is accessible through observation and empirical research as well as clinical and analytical experience. In contrast, knowledge on the *phylogenetic development* is the product of methodological *speculation* and retrospective assumptions, which take existing physiological features and capacities as a point of departure and project back in order to make sense of its generation within an evolutionary timeframe. For example:

> Fónagy speculates that at a certain point of evolution (probably more recently than previously thought) mental states came to be expressed by means of *vocal mimetics*—laryngeal and oral—and their audible products: tonal movements and sound-images. Some clear traces of this remain 'fossilized' in language development. (Fonagy and Target 2007, 434)

This idea coincides with the broader image of human development in contemporary anthropology. In his book *Cultural Origins of Human Cognition* (1999), Michael Tomasello, for example, proceeds under the assumption that "cumulative cultural evolution depends on two processes, innovation and imitation (possibly supplemented by teaching), which must take place over time such that one step in the process enables the next." (39) He regards the capacity of men to perceive the other as an intentional being, similar to themselves, as the biological heritage of *Homo sapiens*. The cognitive and cultural faculties (he mentions language, symbolic representation, gestures, and cognitive achievements like mathematics) are, however, understood as products of learning adopted during ontogenetic development. In his book he discusses only the latter according to findings from widespread empirical research.

The gap between the *time of evolution* (as an object of speculation) and the *time of ontogenetic development* (as an object of empirical studies) can be bridged by the *time of cultural history*. The latter is the time during which the production of cultural and social meaning takes place; it is accessible for research as an object of philological and archaeological investigations, which study the remains, traces and symptoms of earlier and other cultures (Fig. 4.3).

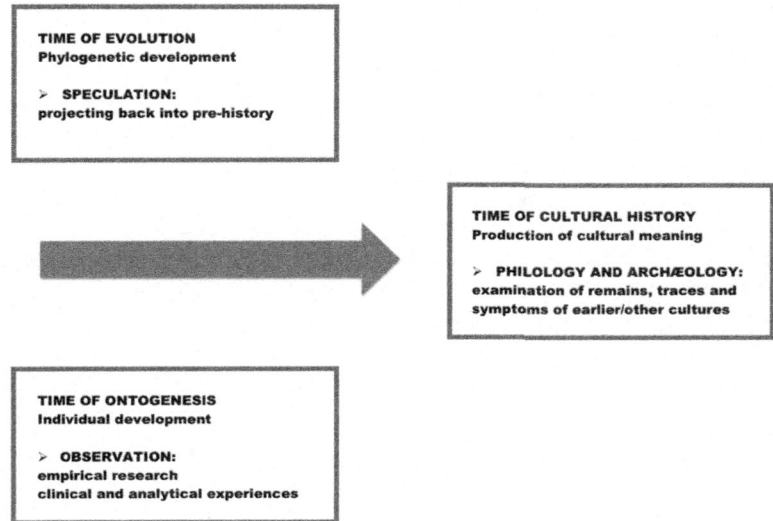

Fig. 4.3 Diagram by the author

In terms of the *coding problem*, the invention of this third, epistemological, temporal perspective is of great importance. It is within this timeframe that a specific human culture forms and takes shape. Concerning the *history of the coding problem*, Walter Benjamin develops an interesting dialectic theory of language in an essay from 1916. His reflections depart from the irreconcilable opposition between the language theories that existed at the time: on the one hand, the idea of similarity between nature/things and words (the "mystical theory of language") and, on the other hand, the idea of an arbitrary language (the "bourgeois theory of language"). Benjamin's theory transforms this opposition into a historical dialectic, arguing that mimetic language has, in the course of the historical development, been subjected and replaced by a language that functions as a system of signs (Benjamin 1996). In a later essay *On the Mimetic Faculty* (1933), he developed this idea further stating that this more recent, conventional system of language (i.e., a system of arbitrary signs) not only functions as a means of communication but simultaneously as a bearer or medium of mimetic elements in language. However, the latter only appears in specific moments or fragmented forms. In his explanation, he refers to different phenomena from childhood as well as past archaic cultures with a mimetic attitude towards nature and the outer world (cult, astrology, dance, macro–micro-cosmos models, etc.). Against this backdrop, he further reflects on the history of the mimetic faculty and states that the gift of being able to perceive similarities is "nothing but a rudiment of the once powerful compulsion to become similar and to behave mimetically" (Benjamin 1999a, 720), i.e., to act according to a mimetic principle. His conclusion reads as follows:

> In this way, language may be seen as the highest level of mimetic behavior and the most complete archive of *nonsensuous similarity*: a medium into which the earlier powers of mimetic production and comprehension have passed without residue, to the point where they have liquidated those of magic. (Ibid., 722; italics added)

By non-sensual similarities he does not mean similarities that are not conceivable via the senses; he refers to a sort of similitude that need not be visually or audibly conceivable because a transferal between the senses and the mind is involved in the operation of similarity itself. Many studies have confirmed Benjamin's interpretation that an overlap of the mimetic system of meaning with conventional (arbitrary) knowledge and language, and an integration of the former into the latter, has taken place during the time of cultural history. For example, Michel Foucault shows in his history of knowledge, *Les Mots et les choses. Une archéologie des sciences humaines* (1966), how *a system of similitude (convenientia, aemulatio, analogia, sympathia)*, which dominated meaning in pre-modern times (one of his main examples is Paracelsus' macro-micro cosmos, Figs. 4.4 and 4.5), has been replaced by a normative, universally valid language system (lexicon, grammar, orthography) during the seventeenth century. The result and indicator of this process is the *Grammaire générale et raisonnée* of Port Royal, published in 1662 (Fig. 4.6). Though this date matters most in French culture, this view on the history of language in general is also valid for other cultures, at least within the European cultural history.

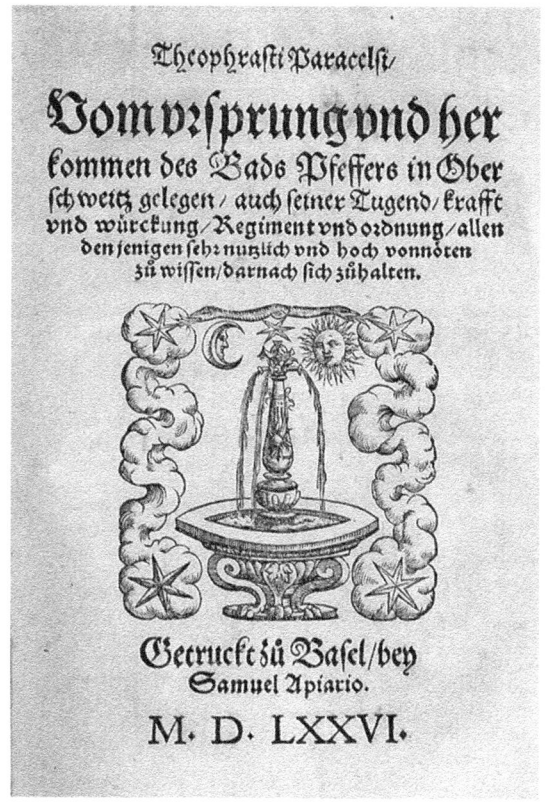

Fig. 4.4 Paracelsus (1576). *Vom Ursprung und Herkommen des Bads Pfeffers in Oberschweitz gelegen* [...]

4 Embodiment in Simulation Theory and Cultural Science, with Remarks… 61

Fig. 4.5 Paracelsus (1493–1541). Woodcut in *Astronomica et Astrologica* (1567)

Fig. 4.6 Antoine Arnauld and Claude Lancelot (1662). *Grammaire générale et raisonnée*

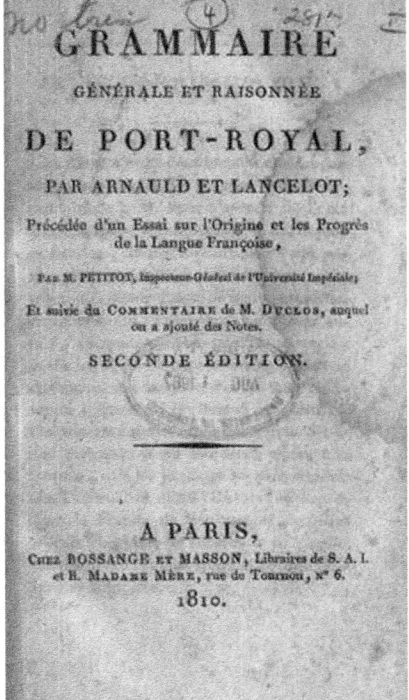

4.6 Language, Image, Gesture

Not only can *Kulturwissenschaft* offer the perspective of historical time as an additional field of investigation that might bridge the gap between evolution and ontogenesis, it also has a large body of knowledge with a differentiated register of expressions at its disposal that might be of interest for simulation theory. As regards the relevance of an *imitation mechanism* in the MNS and *simulation theory*, it might be especially interesting to acknowledge all modes of language within a cultural historical approach as based in an *imagistic notion of language*. This not only means that *similitude* is understood as preceding arbitrary meaning, but that images (in the broader sense, i.e., not only visual images) are conceived as preceding words, or more precisely that the separation of image and word is a product of history. In Europe this separation became prevalent only in the early modern and modern periods, in connection with a conventional system of codes and signs. To present just two definitions of this idea of image, I quote Benjamin's definition from the 1930s and that of William J. T. Mitchell (referring to Foucault) from the 1980s:

> [...I]mage is that wherein what has been comes together in a flash with the now to form a constellation. (*The Arcades Project*, Benjamin 1999a, 463, N3,1)

> The image is the general notion, ramified in various specific similitudes (*convenientia, aemulatio, analogy, sympathy*), that holds the world together with 'figures of knowledge.' (Mitchell 1986, 11)

This theory of image is based in a *material notion of culture* that involves the analysis of the interplay between different bearers and mediators of meaning (bodily enacted, textual, visually reproduced, etc.). In general, *philology and Kulturwissenschaft* can be described as fields that analyze:

- the *production of knowledge and meaning* including both historical and personal perspectives
- *modes of relating*: mimesis, imitation, fiction (from lat. *fictio, fingere*), imagination
- *figurative language*, i.e., tropes, schemata (from gr. σχῆμα), images (in the broader sense), metaphors, visual and verbal embodiments (allegories/personifications)
- the relationship between *denotation* (explicit, intentional meaning) and *connotation* (implicit or involuntary meaning, subtext)
- *kinetic and corporeal expressions*: gestures, expressions of emotion, *pathos formula*, *Gebärden* (from the German word *Gebaren*, to behave)

They are, therefore, not only interested in codes and *rules of signification* but also in all modes of *non-arbitrary meanings* (rhythm of breath, voice, intonation, silence) and in the so-called language of the unconscious (dreams, symptoms, slips).

This whole register of expressions that makes up the object of study for cultural science cannot be classified by separating it into *arbitrary* and *non-arbitrary* parts. Instead each particular mode of producing meaning has to be deciphered in order to identify its conventional elements as well as the connected other part (be it subjected, supplemented, preceding or contradicting) and its quality (whether mimetic, based in similitude, involuntary or unconscious). This means that all human expressions using codes or arbitrary meanings are imbued with other elements, often aspects of similitude or mimesis, with the exception of mathematical nomenclature (although certain figures do hold metaphorical meanings or magic connotations in different cultures).

Needless to say, this approach has a lot in common with psychoanalysis—and, as I want to claim, it has more and more in common with *simulation theory* and the question of *shared meaningful interpersonal space* (Gallese 2009a, 527). And a more intense and differentiated discussion of language in neuro-psychoanalysis that is also more culturally informed might provide the means to close the gap a little further that still exists between the two fields.

4.7 Empathy and Its Relative from Cultural Science: Compassion

In the discussion on simulation theory, the idea of *empathy* as a neuron-based capacity plays a central role. However, it is often unclear whether empathy is regarded just as a basic biological resonance mechanism facilitating a sort of mimetic senso-motoric relationship to the other or whether it is connected with certain ethical values; the latter is the case within the field of socio-neuroscience where empathy is associated with nurturing, care-taking, etc. (Decety and Ickes 2009). In order to demonstrate what the particular relation between simulation theory and cultural history might look like, I will briefly present one example from my own current research. By discussing the chapter of cultural history during which *compassio* emerged, I want to point out how this figure both precedes empathy and overlaps with it.

My thesis is that *embodied simulation*, as an inherited capacity of phylogenesis, provides an innate physiologically based precondition for the development of *intersubjective behavior*, whereas *empathy*, as an emotional relation to others connected with *specific attitudes and values* is the result of history; its production takes place via habituation and a broad variety of *visual-performative modes of embodiment*; here different cultural practices and the arts play an eminent role. That means that *the resonance mechanism of embodied simulation and the different cultural practices interact in the formation of human behavior as it is inscribed and enacted through the body.*

In observing the gestures of human beings, one can detect a great deal about the culture of their ancestors. In this sense, it is not a coincidence that the Hungarian Béla Balázs (1884–1949), the famous theorist of the early stages of film history, became aware, through his examination of gestures in silent film under the heading of Visible Man (1924), that the human body can itself be regarded as an aspect of a 'materialized culture':

> When we see a person's movements or his sensitive hands, do we not recognize the spirit of his ancestors? The fathers' thoughts become the nervous sensitivity, the taste and instinct of the children. Conscious knowledge turns into unconscious sensibility: *it is materialized as culture in the body*. The body's expressiveness is always the latest product of a cultural process. [...] However, although this language of gestures has its traditions, it is unlike grammar in that it lacks strict and binding rules, whose neglect would be severely punished in school. (Balázs 2010, 13–14, transl. mod.)

One consequence of this realization was the idea for a comparative gesturology, a project that is quite similar to what Aby Warburg developed in his *pathos formula picture atlases*, which were collections of the bodily memory of European cultural history handed down by images. My recent research on the history of human fellow feelings, especially on the genesis of compassion, is informed by a similar perspective. Instead of repeating the comprehensive philosophical discourse on pity, empathy and compassion that only tells us what authors thought about how humans should behave, my project undertakes an examination of representations of embodied human behavior, as I analyze sources of a performative culture that came down to us from the past: rituals, pictures, descriptions, music and theatre performances, and prohibitions of certain behaviors, all of which say more about how people actually enacted and performed their affects than reflections by philosophers do.

As for the cultural-historical instances of compassion, the first part of my thesis concerns the concept within the *history of knowledge*: here the figure of *compassio* precedes empathy. As already mentioned, the age of *empathy*, as the canonical narrative of the term in neuroscience has it, reaches back to theories of empathy (*Einfühlung*) in German aesthetic theories around 1900 (Robert Vischer, Theodor Lipps, Wilhelm Worringer). It is sometimes traced back to its so-called 'forerunner' *sympathy*, most notably when the latter concept appeared in the eighteenth century in Adam Smith's *Theory of Moral Sentiments* (1752). However, this version of the figure's past reflects a rather short-term memory of the idea and concept. It has to be grounded in a much longer cultural history that starts with *empatheia* and *sympatheia* in antiquity, followed by its transformation and differentiation into Latin *compassio*, *commiseratio* and *misericordia*, and then to its translation into French *commisération* and *pitié*, English *commiseration* and *pity*, and German *Mitleid* accompanied by *Barmherzigkeit, Erbarmen* and *Mitgefühl*.

The second part of my thesis concerns the historicity of the *human capacity of empathy*. Here, the figure of compassion, in contrast, overlaps with the physiological human prerequisite called empathy. Whereas in neuroscience the MNS and the faculty of empathy are described as a *qualitatively neutral faculty*, the process of embodied simulation and "visuomotor learning" (Gallese 2009a, 530) involve charging empathy with moral and social meaning. Correspondences between phylogenesis and ontogenesis as well as correspondences between cultural history and the process of subjective acculturation are at work here. The particular qualitative sense of empathy, namely a concrete fellow feeling in relation to an other (compassion), is the result of a cultural production that shapes the figure in a shared, meaningful, interpersonal space mediated through visual, gestural and textual images and their respective embodiment. This means that the culturally produced figure plays a role within the process of the so-called *coding* of embodied simulation. As of yet we have no understanding of exactly how the qualitative formation of a specific feeling functions within the MNS. But the history of culture provides a lot of sources for examining the cultural embodiment of *compassion*.

The figure of compassion has been shaped into a powerful embodiment of human attitude through multiple sites of production. Based on the background of my research so far (which includes the history of compassion and empathy but only in European history) it can be said that the attitude of *compassio* was formed:

- as a fellow feeling of an empathic mimetic relationship to another (suffering) person
- as a figure of social behavior within a group or community
- as an indicator for and embodiment of humanity

Although Paul the Apostle wrote long ago in the early years of the founding of Christianity, "Rejoice with them that do rejoice, and weep with them that weep." (Rom. 12.15, *King James Version*), compassion, in the sense of an intersubjective attitude, did not emerge until medieval times in an era when Christianity was transforming from a doctrine and ecclesiastical institution into a social culture, i.e., into a pattern and norm for how a community lives together. It was within this context that *compassion* was formed through multiple scenarios of mimetic and embodied figures. To give just as short an impression as possible in this chapter, I will mention only the most important sites in which the figure of compassion occurred.

A performative site was provided by the Easter or Passion plays (12th/13th cent.) with the invention of the motif the 'lament of Mary' (Fig. 4.7). A vocal-musical site is the *Stabat mater*-song (13th cent.) with the line "Quisest homo qui non fleret" and the musical formula of plorant semiton (weeping halftone) (Fig. 4.8). The *visualization* can be seen in the paintings centered around the motif of the lamenting mother, the iconography of the descent from the cross, the lamentation and the pietà (14th/15th cent.) (Fig. 4.9). Rogier van der Weyden's famous painting *Deposition from the Cross* (c. 1435/40, Museo del Prado Madrid, Fig. 4.10) can be read as a condensed depiction of the embodied simulation of *compassion*. This is obvious not only in the way the collapsing body of the mother resembles the shape of Christ's corpse but also through the group of people surrounding this scene and weeping with Mary. The painting beautifully shows the shift from the theological motif of *passio*, the suffering of a superhuman figure, to the cultural and social constellation of *compassio* as an intersubjective scene, a site of embodied simulation that shapes the idea of compassion (Weigel 2013).

Fig. 4.7 Rogier van der Weyden (1445). *Crucifixion Triptych*

Fig. 4.8 Rogier van der Weyden (1435/40, detail). *Deposition*

Fig. 4.9 Rogier van der Weyden (1435/40, detail). *Deposition*

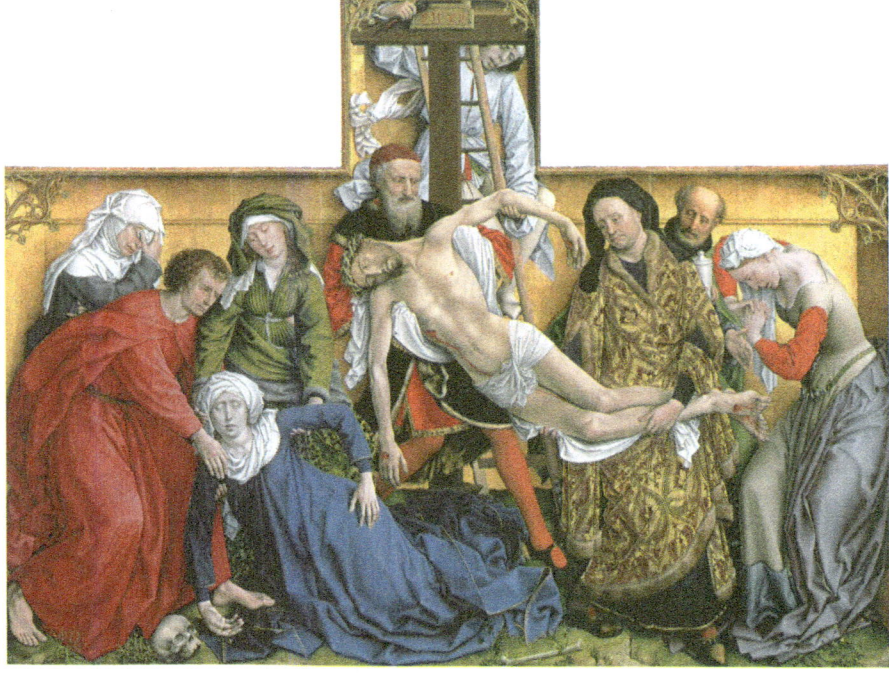

Fig. 4.10 Rogier van der Weyden (1435/40). *Deposition*. Centre piece of a triptych, painted for a church in Leuven, today in the Prado (Madrid)

In this historical constellation *compassion* emerges as a basic requisite not only for the individual but also for an intersubjective attitude and behavior. Hence it is also a precondition for the constitution of a community. Compassion is a product of a constellation that combines (1) a mimetic intersubjective relationship, a *dual* configuration of intercorporeality (the observer in relation to Mary/the mother), and (2) a meaningful *space* in which Mary functions as the mediator to a higher moral meaning. Together this forms the site of a *triadic constellation* (Fig. 4.11) that includes the threshold between the dyadic configuration of a corporeal relation to the other and a superimposed symbolic system.

Since this constellation, namely, the threshold between a dyadic structure and a triadic one, resembles the basic, conflict-laden scenario in psychoanalysis, it is interesting for a case study of the relation of empathy and compassion, or in terms of epistemology between neuroscience and *Kulturwissenschaft*. Moreover, the constellation refers to controversial interpretations within the field of psychoanalysis concerning the different stages within a subject's development, especially the transition from the relationship to the (body of the) mother/other to the entry into social patterns (in Lacanian terms: the transition from the imaginary to the symbolic).

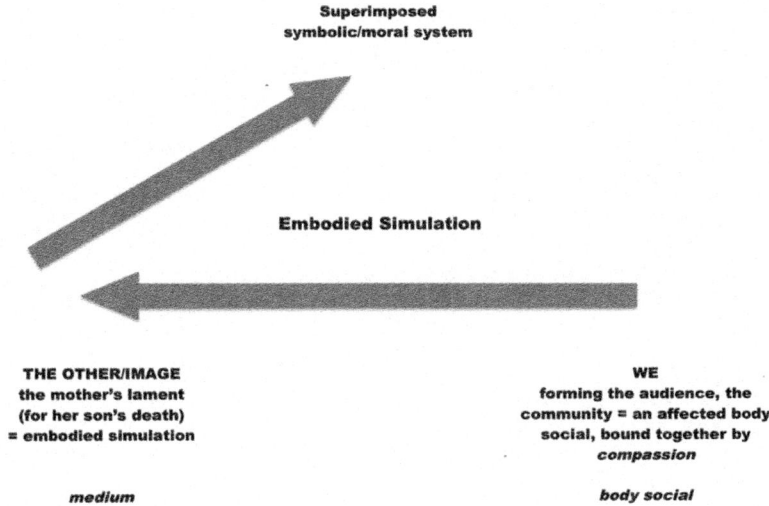

Fig. 4.11 Diagram by the author

References

Arendt, H. (2002). *Denktagebuch*. München/Zürich: Piper.
Balázs, B. (2010). *Early film theory: Visible man and the spirit of film*. New York: Berghan Books.
Benjamin, W. (1996). On language as such and on the language of men. In M. Bullock & M. W. Jennings (Eds.), *Selected writings* (Vol. 1.1913–1926, pp. 62–74). Cambridge, MA/London: Harvard University Press.
Benjamin, W. (1999a). On the mimetic faculty. In M. W. Jennings, H. Eiland, & G. Smith (Eds.), *Selected writings* (Vol. 2.1927–1934, pp. 720–722) Cambridge, MA/London: Harvard University Press.
Benjamin, W. (1999b). *Arcades project* (H. Eiland & K. McLaughlin, Trans.). Cambridge, MA/London: Harvard University Press.
Benjamin, W. (2002). *Berlin childhood around 1900*. In M. W. Jennings & H. Eiland (Eds.), *Selected writings* (Vol. 3.1935–1938, pp. 344–413). Cambridge, MA/London: Harvard University Press.
Blumenberg, H. (1996). *Shipwreck with spectator. Paradigm of a metaphor for existence* (S. Rendall, Trans.). Cambridge, MA: The MIT Press.
Blumenberg, H. (2007). *Theorie der Unbegrifflichkeit*. Frankfurt am Main: Suhrkamp.
Blumenberg, H. (2010). *Paradigms for a Metaphorology*. Ithaca: Cornell University Press.
Craig, A. D. "Bud". (2011). *How do you feel? The neuroanatomical basis for human awareness of feelings from the body*. Paper at the 12th international npsa-congress, Berlin.
Damasio, A. (1999). *The feeling of what happens: Body and emotion in the making of consciousness*. London: William Heinemann.
Damasio, A. (2010). *Self comes to mind: Constructing the conscious brain*. New York: Pantheon Books.

Decety, J., & Ickes, W. (Eds.). (2009). *The social neuroscience of empathy*. Cambridge: The MIT Press.
Dilthey, W. (1990). *Einleitung in die Geisteswissenschaften. Versuch einer Grundlegung für das Studium der Gesellschaft und der Geschichte*. Göttingen: Vandenhoeck&Ruprecht.
Fonagy, P., & Target, M. (1996). Playing with reality: I. Theory of mind and the normal development of psychic reality. *International Journal of Psycho-Analysis, 77*, 217–233.
Fonagy, P., & Target, M. (2007). The rooting of the mind in the body: New links between attachment theory and psychoanalytic thought. *Journal of the American Psychoanalytic Association, 55*(2), 411–456.
Fonagy, P., Gergely, G., Jurist, E., & Target, M. (2002). *Affect regulation, mentalization and the development of the self*. New York: Other Press.
Foucault, M. (1966). *Les Mots et les choses. Une archéologie des sciences humaines*. Paris: Gallimard.
Freud, S. (1895). *Project for a scientific psychology*. S.E. I, 281–397.
Freud, S. (1900). *The interpretation of dreams*. S.E. IV/V, 1–627.
Freud, S. (1915). *Instincts and their Vicissitudes*. S.E. XIV, 109–140.
Freud, S. (1920). *Beyond the pleasure principle*. S.E. XVIII, 1–64.
Goldschmidt, G.-A. (2000). *Quand Freud voit la mer. Freud et la langue allemande*. Paris: Buchet/Chastel.
Gallese, V. (2009a). Mirror neurons, embodied simulation, and the neural basis of social identification. *Psychoanalytic Dialogues, 19*(5), 519–536.
Gallese, V. (2009b). We-ness, embodied simulation, and psychoanalysis: Reply to commentaries. *Psychoanalytic Dialogues, 19*(5), 580–584.
Gallese, V. (2011). *Bodily selves in relation: Embodied Simulation and inter subjectivity*. Paper at the 11th international NPSA-congress, New York.
Gallese, V., & Di Dio, C. (2011). Neuroesthetics: The body of esthetic experience. In V. Ramachandran (Ed.), *Encyclopedia of human behavior* (2nd ed.). San Diego: Academic.
Kaplan-Solms, K., & Solms, M. (2000). *Clinical studies in neuro-psychoanalysis: Introduction to a depth neuropsychology*. New York: Karnac Books.
Kristeva, J. (1984). *Revolution in poetic language*. New York: Columbia University Press.
Lakoff, G. (1997). How unconscious metaphorical thought shapes dreams. In D. J. Stein (Ed.), *Cognitive science and the unconscious* (pp. 89–120). Washington, DC: American Psychiatric Press.
Lakoff, G., & Johnson, M. (1999). *Philosophy in the flesh: The embodied mind and its challenge to Western thought*. New York: Basic Books.
Lux, V. & Weigel, S. (Eds.). (2015). *Empathy. A neurobiological capacity and its cultural and conceptual history*. London, New York: Palgrave Macmillan (forthcoming)
Mancia, M. (2006). *Psychoanalysis and Neuroscience*. Berlin: Springer.
Mitchell, W. J. T. (1986). *Iconology – Image, text, ideology*. Chicago/London: University of Chicago Press.
Moritz, K. P. (1968). *Über die bildende Nachahmung des Schönen*. Nendeln/Liechtenstein: Kraus (Original work published 1888).
Panksepp, J. (1998). *Affective neuroscience: The foundations of human and animal emotions*. Oxford: Oxford University Press.
Papapetros, S. (2012). *On the animation of the inorganic*. Chicago: University of Chicago Press.
Plessner, H. (1965). *Die Stufen des Organischen und der Mensch. Einleitung in die philosophische Anthropologie*. Berlin: De Gruyter.
Plessner, H. (2003). *Anthropologie der Sinne. Gesammelte Schriften in zehn Bänden* (Vol. 3). Frankfurt am Main: Suhrkamp.
Rizzolatti, G. & Sinigaglia, C. (2008). *Mirrors in the brain. How our minds share actions and emotions* (F. Anderson, Trans.). Oxford: Oxford University Press. (ital. original 2006: *So quel que fai, il cervello che agisce e i neuroni specchio*)

Shimada, S., & Hiraki, K. (2006). Infant's brain responses to live and televised action. *NeuroImage, 32*(2), 930–939.

Simmel, G. (1992). *Soziologie. Untersuchungen über die Formen der Vergesellschaftung*. Frankfurt am Main: Suhrkamp.

Solms, M. (2000). Dreaming and REM sleep are controlled by different brain mechanisms. *Behavioral and Brain Sciences, 23*, 843–850.

Solms, M. (2001). The interpretation of dreams and the neurosciences. *Psychoanalysis and History, 3*, 79–91.

Solms, M., & Saling, M. (1986). On psychoanalysis and neuroscience: Freud's attitude to the localizationist tradition. *The International Journal of Psycho-Analysis, 67*, 397–416.

Solms, M., & Oliver, T. (2002). *The brain and the inner world: An introduction to the neuroscience of subjective experience*. New York: Other Press/Karnac Books.

Tomasello, M. (1999). *Cultural origins of human cognition*. Cambridge, MA: Harvard University Press.

Turnbull, O. (2001). The neuropsychology that would have interested Freud most. *Neuro-Psychoanalysis, 3*(1), 33–38.

Warburg, A. (2010). *Werke in einem Band*. (M Treml, S. Weigel, & P. Ladwig, Eds.). Berlin: Suhrkamp.

Weigel, S. (2004). Phantombilder zwischen Messen und Deuten. Bilder von Hirn und Gesicht in den Instrumentarien empirischer Forschung von Psychologie und Neurowissenschaft. In B. von Jagow & F. Steger (Eds.), *Repräsentationen. Medizin und Ethik in Literatur und Kunst der Moderne* (pp. 159–198). Heidelberg: Winter.

Weigel, S. (2006). *Genea-Logik. Generation, Tradition und Evolution zwischen Kultur- und Naturwissenschaften*. Munich: Wilhelm Fink.

Weigel, S. (2007). The measurement of angels. Images of pure mind as a matter of science in 19th century. *European Review, 15*(3), 297–320.

Weigel, S. (2013). Tränen im Gesicht. Zur Ikonologie der Tränen in einer vergleichenden Kulturgeschichte von Trauergebärden. In S. Weigel (Ed.), *Gesichter. Kulturgeschichtliche Szenen aus der Arbeit am Bildnis des Menschen* (pp. 103–126). Munich: Fink.

Weigel, S. (2015). *Grammatologie der Bilder*. Berlin: Suhrkamp.

Part III
The Unconscious Before Freud and After

Chapter 5
Signs and Souls: The Prehistory of Psychoanalytical Treatment in Nineteenth-Century French Psychiatry

Gerhard Scharbert

Abstract The chapter traces three "generations" of French psychiatrists from the nineteenth century, whose work prepared the grounds for Freud's step from neurology to questions of language and other signs. (1) Other than Franz Joseph Gall, who at the time investigated the organic causes of mental diseases, Philippe Pinel (1745–1826), one of the founders of the concept of psychiatric 'analysis', concentrated on results obtained from analytical situations: namely, observations of patient behaviour and the statistical compilation of data. (2) Three students of Pinel: Xavier Bichat (1771–1802), best known for his studies on human tissue, who established a link between the 'signs' and 'functions' of a mental disease and their organic correlate in the body. Victor Broussais (1772–1838), who established a theory of the interdependence of the various bodily organs, and who postulated that the concepts of "reason, ego, consciousness are only expressive of the results of actions of encephalic nervous matter; action that may change constantly throughout life." And Dominic Esquirol (1772–1840), who furthered Pinel's idea of clinical psychiatry. (3) The essay's historic trajectory concludes with Jacques-Joseph Moreau de Tours (1804–1884), a student of Esquirol, who experimented systematically with Cannabis and explored the effects of drugs on the nervous system, establishing a link between the various forms of deliria, whether drug induced or classified as mental illness, and thus "navigated on the frontier between psychiatry and neurology."

Keywords Psychiatry • Neurology • Organic disease • Language • Signs

Certain essential aspects of the psychoanalytic method of treatment are rooted in the preliminary theoretical work and hypotheses of nineteenth-century French psychiatry. Due to his knowledge of this preliminary work, which he was able to deepen during his research stint with Charcot in Paris, Sigmund Freud was able to situate

---

G. Scharbert (✉)
Institut für Kulturwissenschaft, Philosophische Fakultät III, Humboldt-Universität zu Berlin,
Unter den Linden 6, D-10099 Berlin, Germany
e-mail: gerhard.scharbert@hu-berlin.de

© Springer International Publishing Switzerland 2016
S. Weigel, G. Scharbert (eds.), *A Neuro-Psychoanalytical Dialogue for Bridging Freud and the Neurosciences*, DOI 10.1007/978-3-319-17605-5_5

his own neurological findings in pediatric clinical neurology[1] and theoretical aphasia research[2] in a new and dynamic[3] context. In this context, linguistic signs, which had previously only been considered as symptoms by his predecessors and contemporaries in psychiatry and neurology, would come to play a crucial role, not only in etiology but also in therapy (namely, as the talking cure). Even some of Freud's central concepts, including that of psychoanalysis itself, can be traced back in their inception to Philippe Pinel (1745–1826).

Pinel, the head of the psychiatric institution Salpêtrière in Paris, and in this function Jean-Martin Charcot's predecessor, published his *Traité médico-philosophique sur l'aliénation mentale ou la manie* in 1801. In the introduction, he acknowledges the work of the Scottish physician Alexander Crichton (1763–1856), whose psychopathology of emotions had made him famous beyond the borders of his own country.[4] The concept of analysis in Pinel's introduction, which would come to play an important role in French psychiatry of the nineteenth century, is also indebted to him:

> Crighton [sic] semble s'être élevé à un point de vu étendu que ne peuvent atteindre le métaphysicien et le moraliste, c'est la considération des passions humaines regardées comme des simples phénomènes de l'économie animale, sans aucune idée de moralité ou d'immoralité, et dans leur rapports simples avec les principes constitutifs de notre être, sur lesquels elles peuvent exercer des effets salutaires ou nuisibles. [...]
>
> Il a soumis à une sorte d'analyse le principe de nos actions, et il en a trouvé la source dans les penchans primitifs qui dérivent de notre structure organique. (Pinel 1801, xxii, xxxix f.)

Following Crichton, Pinel also declares the obsolescence of hypotheses about human nature and understanding;[5] far more important, he argues, are the meticulous patient observations and the statistical assessment of findings (Pinel 1801, xlvii f., tables on 109 and 250). His interest in the precise organic causes of those afflictions, whose external features he so minutely describes, is strikingly limited. Pinel rejects explanations that seek causes, for example, in brain anatomy because he considers insanity to be a nervous disorder instead. In this respect he follows another Scottish physician, William Cullen (1710–1790), who coined the term *neurosis* and

---

[1] Among the numerous publications and reviews concerning this topic, I would like to mention here the representative text, Sigmund Freud's *Klinische Studie über die halbseitige Cerebrallähmung der Kinder* (1891a).

[2] See Freud's *On Aphasia/Zur Auffassung der Aphasien* (1891b).

[3] The term "dynamic psychiatry" was coined by Henri F. Ellenberger in his work "The Discovery of the Unconscious: The History and Evolution of Dynamic Psychiatry" (1970) in order to designate the opposition between the anatomical-localizationist and the dynamic standpoints in psychiatry.

[4] On Crichtons work, see Gerhard Scharbert (2010, 49 ff.).

[5] "Tout cet ensemble de faits peut-il se concilier avec l'opinion d'un siège ou principe unique et indivisible de l'entendement? Que deviennent alors des milliers de volumes sur la métaphysique?" (Pinel 1801, 25).

determined "that almost all diseases could be called *nervous diseases*." (Cullen 1785, 182)[6]

Pinel was not at all discouraged by the bewildering multifariousness of phenomena, for he was convinced that he could put them in an operational order through precise observation of the symptoms and courses of the diseases. This would serve him as the basis for a diagnosis and possible treatment of the patient:

> Veut-on tracer et décrire les phénomènes de l'aliénation mentale, c'est-à-dire d'une lésion quelconque dans les facultés intellectuelles et affectives, on (1) ne voit que confusion et désordre, on ne saisit que des traits fugitifs qui n'éclairent un moment que pour laisser ensuite dans une obscurité plus profonde, si on ne part comme d'un terme fixe de l'analyse des fonctions de l'entendement humain. (Pinel 1801, 1 f.)

Pinel, who was known not simply as a "doctor for the insane" but primarily as a specialist for internal medicine,[7] used these precepts to transfer the methodological foundation of the clinic (namely the precise observation of patients, the minute description of the courses of illnesses, and the unity of research and therapy at the patient's bedside) from internal medicine to the clinical psychiatry that he helped to found.

It is at this juncture that one of the central problems of early psychiatry emerges: If internal medicine had succeeded in developing a systematic symptomatology, which referred to the course of illnesses and the affected organs or tissues, then psychiatry's premise of "nervous disorders" opened a new realm of knowledge; but by turning away from topographical explanations, it constitutes its object of study as an organic ailment without a distinct physical location, as an illness which is only recognizable through its external signs and only describable in its dynamics. As a consequence, Pinel's *Traité* ascribes the highest value to the doctor's exacting gaze, who observes the signs of the illness and its course; the temporal form of pathological occurrences thereby gains central importance (Pinel 1801, lii f.). The diagnostics of the physician's gaze observes and analyzes, while the differentiation of symptoms and the numerical compilation of patient histories inexorably supplant etiology, prognostics, and therapy. "Madness" disintegrates into a classification of external signs, whose structure is reproduced by the medical language that precisely analyzes it based on the greatest possible number of physical manifestations (ibid., xxxix).

Pinel himself introduced his concept of analysis to the medical profession in two texts: with *Nosographie philosophique ou la méthode de l'analyse appliquée à la médecine* (1798) and *La médecine rendue plus précise et plus exacte par l'application de l'analyse* (1802), both published in Paris. As a medical student in Toulouse, he had already taken up the study of mathematics. The topic of the study with which he presumably earned the degree of *Baccalaureus in medicina* in 1773 (see Lechler 1960, 95 ff., 112), *De la certitude que l'étude des mathématiques imprime au*

---

[6] "Quantum ego quidem video, motus morbosi fere omnes a motibus in systemate nervorum ita pendent, ut morbi fere omnes quodammodo Nervosi dici queant."

[7] Through his famous *Nosographie philosophique ou la méthode de l'analyse appliquée à la medicine* (Pinel 1798).

jugement dans son application aux sciences, alludes to the significance of exact methods that he would subsequently employ in later works. Up until his time in Paris, Pinel financed his studies through private tutoring in mathematics.[8] His concept of analysis, which occupies a central position in his medical views, is formed by a practically oriented gaze in the sciences of the eighteenth century. "Analytic thought of the time [...] was also reflected in the works of mathematicians. One began to systematically analyze the problems of physics and engineering with the means attained by Newton and Leibniz. A particular working method emerged from differential and integral calculus, which one still calls analysis today" (Popp 1981, 76 ff.). Pinel wrote retrospectively about his beginnings:

> I am among the proponents of strict observation, and I limit myself to pointing out that the method, which has been demonstrated and developed in my works, is the fruit of many years of preliminary study and the practice of medicine in large hospitals; [...] This constitutes a type of experiment that began in 1774 according to a precise plan and is still ongoing [...]. (Dictionnaire des Sciences Médicales Paris 1812, 204; quoted after Lechler 1960, 118)[9]

But what does it mean to conduct an experiment in a psychiatric field that is concerned, in Pinel's own words, with "external features and [...] physical transformations," which "could correspond to the functions of the understanding or of the will"? (Pinel 1801, xlii)

A clinical practice whose object remains in organic obscurity must concentrate in essence on the analysis of symptoms and their classification. The correspondences between diseased functions and bodily, linguistic, or written forms of expression are thus the product of a highly specific medical gaze. This gaze beholds madness's exterior. The doctor can only be assured of his subject matter when he maintains the experimental nature of his own work as a permanent condition. Because the attribution of signs and illnesses only appears to be possible through a large number of cases in the first place, the doctor's analysis is also no longer bound to an operating table or a sick bed.

Analysis, defined by Pinel as based on observation, can be combined with statistical analysis—as would become characteristic of his psychiatric work. But it also includes fragmenting description, the minute notation of manifestations and processes and their generalization as types, basic forms, and schemas. While the philosopher and physician Pierre-Jean-Georges Cabanis (1757–1808), a friend and patron of Pinel, had already attempted to answer the virulent question of certitude (first raised by the *corpus hippocraticum*) on a new level by recommending a closer connection to logic as well as the conceptual rigor of philosophy to physicians in his *Du degré de certitude de la médicine* (1798), Pinel provided certitude for the difficult field of psychiatry—in accordance with the established model of the natural sciences of

---

[8] "[...] quant à ma situation actuelle a Paris elle est aussi agréable que je pouvais l'attendre; comme les leçons de mathematiques sont beaucoup mieux payées ici qu' en province je me procure une honnête aisance [...]." Letter by Pinel to his brother, dated 8.12.1778, quoted after Lechler 1960, 145.
[9] Wrongly dated from 1821 in the references of Lechler.

physics, chemistry, and botany (Pinel 1801, xl ff.)—by postulating a common, yet unspecific organic basis, ascertaining respective appearances, conducting statistical evaluations, and finally conveying all those elements in a coherent conceptual order.

Psychic illness thus dutifully appeared for a brief moment as an act in the theater of this revolutionary philanthropy, which Pinel had so momentously staged with his high-profile liberation of the insane from the real chains of the *ancién régime*, only to immediately disappear again behind heaps of numbers used to record the dark glances, incomprehensible talking, and impenetrable skulls of patients—as well as behind the walls of the new clinics (see Foucault 2006).

The problem remained that the psychiatric analysis developed by Pinel, a configuration of vague neurophysiological references with multifaceted, elaborately classifiable visibility, was not capable of developing the space of illness to as subtle a degree as had been possible in internal medicine following the discovery and classification of tissue by Xavier Bichat. In the field of internal medicine, Pinel had also initially earned his fame among contemporaries through the noted *Nosographie philosophique*, and yet here too his celebrity status would fade due precisely to other new knowledge. Since he made no effort to conceal his disdain for brain anatomy in his *Traité*, which also set him against the celebrated brain anatomist Franz Joseph Gall,[10] it was left to his students to exploit this terrain for French psychiatry. The differences that arose from Pinel's clinical approach to external signs of madness, on the one hand, and Gall's anatomically grounded organology supported by findings in histology, on the other hand, prevented the two from taking notice of one another, despite all apparent affinities. Gall's studies of the brain and nervous system and the conclusions he drew from them for psychiatry in particular constituted a scientific tool to be used for precisely those problems, which Pinel's analysis were not able to address. In the wake of sensualist views and given the form that Cabanis lent them as corresponding to a physiology of sensibility and irritability, Pinel proceeded "from the analysis of cognitive performance as from a fixed point," which, to be sure, could have "effects on the animalist economy," yet in essence was traced back to external impressions and their mediation by the senses (Pinel 1801, xlv). In his *Untersuchungen über die Anatomie des Nervensystems überhaupt, und des Gehirns insbesondere. Ein dem französischen Institute überreichtes Mémoire* (1809), Gall responded to this with his organology of the brain and nervous system as the anatomical basis for all abilities and combined sensations, which posed a reversal to the physiological sequence. Where one had previously only seen forms and mechanical connections, the entire material effort of all activities of the soul was now revealed. Just as the different viscera are controlled by single and particular nervous systems and just as the different activities of the senses happen through specific individual nerves, so too are the different proclivities and abilities of the animal world and of human beings caused by means of specific individual parts of the brain (quoted after Lesky 1979, 72).

---

[10] As a member of the committee of the *Académie des Sciences,* in charge of the examination of Gall's thesis of 1808, Pinel damaged Gall's reputation in France considerably.

Beyond his phrenology, which, as a form of "universal human knowledge," had a similar reception among the broader public as Lavater's physiognomy, one must regard Gall's theories and results as a crucial mediation of clinical-analytic and comparative-experimental methods within French psychiatry in particular. His postulate, in *Sur les fonctions du cerveau et sur celles de chacun de ses parties* (1822), that "mental illnesses are illnesses of the brain" (quoted in Lesky 1979, 134) ushered in a development that opened new fields of knowledge pertaining to psychic illnesses, such as pharmacology and genetics.

Philippe Pinel's most important students, Xavier Bichat, Victor Broussais, and Dominique Esquirol, worked in the areas of physiology, pathological anatomy, and psychiatry. Hence they represented realms of knowledge at whose crossroads the material would further be handled in the first half of the nineteenth century, realms of knowledge brought together from internal medicine and psychiatry by their teacher with a new focus: the clinic. The part of Pinel's medical thought that primarily aimed at nosographic classification of disorders and which had his training in Montpellier as well as the systematics of William Cullen to thank for central impulses underwent a significant shift through the work of his students.

Xavier Bichat's point of departure was anatomy and surgery. Since the founding of anatomical pathology by Giovanni Battista Morgagni (1682–1771), pathological aspects of anatomy had receded behind either physiological or clinical attempts at description. Names like Albrecht von Haller, Théophile de Bordeu and, above all, Pinel himself each represent a specific view of the human as organism, according to which the phenomena of life stand in relation to signs of diseases that are only indirectly graspable. Whereas physiologically inspired observation brought to light fundamental elements of biological functions, indeed of the functional in general, the clinic confronted it with a plethora of images of illness and their courses, which could ultimately lead to death; their connection to the affected structures, however, remained to a great extent unclear. Bichat brought these elements together on the basis of an anatomy that operates physiologically and implements questions of pathology experimentally. He applied Pinel's *méthode de l'analyse* not to the form of diseases, that is, to a temporal space which merely intersected with corporeal functions, but to the space of the body in its very construction; in this way, he arrived at a description of organic structures that produced a direct connection between physiological functions and their impairments due to illness. His physiology formulated possibilities for substantiating these classifications through experimentation.

Victor Broussais (1821), on the other hand, presented the clinical views of his teacher in the field of psychiatry with a neurologically oriented, general clarification of causality grouped around the central concept of *irritation* and the connected *sympathies morbides* (Broussais 1821, xxii) and *relations* (ibid., xxxiii): "L'irritation peut exister dans un système, sans qu'aucun autre y participe; [...] Les nerfs sont les seuls agents de l'irritation [...]" (ibid., xxxii). Against the background of Broussais' view that an irritant inflammatory process can be regarded as the cause of all illnesses, *sympathy* has to be understood here in the sense of general affection. The concept of "relations" (ibid., xxxiii) in reference to the diseases that befall Bichat's "animalistic characteristics" of life (i.e., the organic-physiological functions, in this case, of the nervous system) implies a construction that renders "nervous disorders" visible as a form of expression of hidden cerebral processes. This facilitates the connection to the anthropological ambitions hinted at by Gall, which Broussais follows in essential points. Reason, the self [*moi*], and consciousness are for Broussais only an expression of the results of the actions of nervous brain matter, which change over the entire course of a lifetime: "[…] les mots raison, moi, conscience, n'expriment que des résultats de l'action de la matière nerveuse de l'encéphale; action qui est susceptible de changer tant que dure l'état de vie" (Broussais 1828, 490).

For madness in particular, this view entails a classification of psychic ailments within the realm of *dégénérescence*, of degeneration, which Jean Étienne Dominique Esquirol (1772–1840), the third of the above named students of Pinel and psychiatrist at the Salpêtrière, described as the organic basis of *maladies mentales* (Esquirol 1838). He went on to connect degeneration with the cultural space of modern Paris. Broussais writes: "La manie suppose toujours une irritation du cerveau: cette irritation peut y être entretenue long-temps par une autre inflammation, et disparaître avec elle; mais si elle se prolonge, elle finit toujours par se convertir en une véritable encéphalite, soit parenchymateuse, soit membraneuse." (Broussais 1821, xxx) *Irritation* of the brain entails an increase of innervation; the hypertrophy of certain features already prefigured in Gall are, in other words, relocated in the nervous-pathological realm because nerves are indeed the sole agents of the transmission of irritation. On the basis of this explanation, Jacques-Joseph Moreau de Tours (1804–1884) would later update the *analysis* of personality functions formulated by Broussais with his concepts of *état nerveux héréditaire* and *idiosyncrasique* (Moreau de Tours 1859, 429).

In the second edition of his major two-volume work *De l'irritation et de la folie*, Broussais names a factor that Pinel had already identified as essential for his *methode d'analyse*. Proceeding from the question of what counts as an object of psychology in the first place, he arrives at the analysis of those signs that first constitute a genuine object of analysis for the psychologist: linguistic signs. Under the heading of *valeur des signes*, the psycho-physiologist[11] tests the suitability of linguistic signs as an object of medical analysis (Broussais 1821, 207 ff.). He comes to a conclusion that supplements his findings from the first edition of his book in a characteristic manner.

> Il est certain que la haute intelligence tient aux organes phrénologiques de la comparaison et de la causalité, puisqu'elle est toujours en raison du développement simultané de ces deux organes; mais cherchons par quoi elle se manifeste.
>
> Evidemment c'est par le langage, soit parlé, soit écrit; […] Nous avons la faculté, 1° de représenter aux autres hommes, par des signes, un objet qui a frappé un de leurs sens; 2° de leur faire comprendre et éprouver ce que nous avons senti quand cet objet a frappé nos sens, et certes nous avons pu sentir bien diversement.

---

[11] For the contemporaneous understanding of this concept, see, for instance, Johannes Müller (1822) or Pierre-Jean-Georges Cabanis (1799, 1).

[...] Il y a donc deux ordres de signes: 1° les uns qui représentent les objets extérieurs; 2° les autres qui représentent notre sentiment personnel modifié par les objets. (Ibid., 531 f.)

This definition of language already contains the mandate for later experiments that would impact the brain: Language, whether spoken or written, renders audible or legible what plays out in the invisible physiology of the nervous system; the physiologist merely has to understand how to trace linguistic signs to their foundation in the changes of nervous brain mass.

Finally, Jean Étienne Dominique Esquirol, Pinel's student and later the teacher of Jacques-Joseph Moreau de Tours, further developed the elements of Pinel's clinical psychiatry in subtle ways; and he did so in closer connection to his teacher than did Bichat and Broussais, who made their mark in internal medicine and physiology. And yet, faced with the altered situation in France, Esquirol gave a palpably unique accent to Pinel's penchant for nosographic analysis and statistical assessment of cases. Like Broussais, he was indebted to Gall (and not to Pinel) for the organological foundation of his psychiatry.

The lapidary definition of his object of study in the first chapter of his book *De maladies mentales considérées sous les rapports médical, hygiénique et médico-légal* resounds with clear reverence for Gall and a certain affinity to Broussais: "La manie est une affection cérébrale, chronique, ordinairement sans fièvre, caractérisée par la pertubation et l'exaltation de la sensibilité de l'intelligence et de la volonté." (Esquirol 1838, 132) The problems that have informed discussions of psychopathology since the eighteenth century are reiterated in this sentence, compressed once again in *the* form with which the nascent psychiatry of France at the beginning of the nineteenth century had taken them up and systematized them: Psychic illness is a *mental* illness (*maladie mentale*), in other words, a disease of the intelligible abilities in their organic basis; it is chronic, slowly progressing, and difficult to cure; it is not inflammatory, but precedes *affection* and *sensibility* in interaction with the nervous system.[12] Through "disorders [...] of the activity of understanding and the will," it withdraws from the domain of internal medicine and physiology, at least with regard to their operative methods.

This is where, in a remarkable synthesis of experiment and analysis, the work of a student of Esquirol begins. Jacques-Joseph Moreau de Tours all of a sudden saw crystal clear the interplay of linguistic signs and the organic influence on the life of the soul. He could thereby draw upon the preliminary work of François Magendie (1783–1855), a pioneer of experimental pharmaco- and neurophysiology, which for its part was indebted to the colonial conquest of the plant world of the Orient, which secured France a leading role in the analysis of pharmaceutical drugs in the nineteenth century. Magendie concisely formulated the principle of his research in the sentence: "La pathologie du système nerveux n'est donc autre chose que la physiologie expérimentale appliquée à l'homme." (Magendie 1841, 172)

In his book, *Du hachisch et de l'aliénation mentale. Études psychologiques*, published in 1845, Moreau de Tours introduces the idea of a principle unity of

---

[12] On the discussion of fever see Michel Foucault (1973), Chapter X.

délire, a concept which he elaborates in the first part of this text using written material from patient reports, in order to corroborate his experimental approach, which aims to lead him to the hidden source of *folie*, his collective term used to subsume several other concepts (Moreau de Tours 1845, 30).[13]

> J'avais vu dans le hachisch, ou plutôt dans son action sur les facultés morales, un moyen puissant, unique, d'exploration en matière de pathogénie mentale; je m'étais persuadé que par elle on devait pouvoir être initié aux mystères de l'aliénation, remonter à la source cachée de ces désordres si nombreux, si variés, si étranges qu'on a l'habitude de désigner sous le nom collectif de *folie*. (Ibid., 29 f., emphasis in the original)

The medium of this expedition into the territory of mental pathologies is self-observation, *observation intérieure* (ibid., 30), which no longer draws its conclusions from a higher, philosophical vantage point but, in contrast, from the conditions of an artificially induced confusion and dissolution of boundaries. Moreau de Tours thus responds to an argument from the contemporary philosophical critique of psychiatric methods as such, and he reacts in a way fundamentally appropriate to his discipline.

Auguste Comte claimed that all reflexive self-observation is nothing more than "pure illusion" (Comte 1864, 31).[14] Moreau then skillfully transformed the much criticized weakness of philosophical reflection into a weapon in the dispute between philosophy and psychiatry over which discipline had priority by wielding it against its creator and distinguishing the character of experimental psychiatric knowledge from the modes of knowledge criticized by Comte: The site of *observation intérieure* is not the *domaine d'une nuageuse métaphysique*, but rather the observations every individual might be able to conduct upon him or herself, for example by dreaming (Moreau de Tours 1845, 30 f.). And through the use of hashish, even conditions related to madness can be experienced in a reversible, non-permanent way. Because they could later be recorded by the very person who experienced the effects, such conditions could become an object of the empirical or experimental sciences: "De cette manière, et guidé exclusivement par l'observation, mais par ce genre d'observation qui ne relève que de la conscience ou du *sens intime*, j'ai cru pouvoir remonter à la source primitive de tout phénomène fondamental du délire" (ibid., 31, my emphasis).

The *sens intime* to which Moreau refers here is invoked as a mode of knowledge that is fundamentally distinct from the philosophical introspection of the intellect. Comte criticized this aspect as part of his founding of the new science of sociology. Moreau's *sens intime* is a physiological sense that reports to the subject what is presently occurring in the brain, even when the report is merely of an unholy confusion induced by drugs or the memory of a particularly bizarre dream. Indeed, this sense enables something to be seen that would otherwise go unnoticed by the eyes, particularly the sociological and philosophical ones.

---

[13] Moreau emphasizes here that the terms *délire*, *folie*, and *aliénation mentale* are interchangeable.

[14] In *Cours de philosophie positive* (1830) Comte spoke of two forms of observation, "(…) l'une extérieure, l'autre intérieure, et dont la dernière est uniquement destinée à l'étude des phénomènes intellectuels," and asserted, "(…) cette contemplation directe de l'ésprit par lui-même est ne pure illusion."

> [...] j'ai dû admettre, pour le délire en général, une nature psychologique, non pas seulement analogue, mais *absolument identique* avec celle de l'état de rêve.
>
> Cette identité de nature qui échappe à l'observation extérieure, c'est-à-dire qui ne s'exerce que sur autrui, est clairement constatée, je puis dire perçue par l'observation intime. (Ibid., emphasis in the original)

Moreau's remarks imply a critique of contemporary psychoanatomy, which cannot produce any results that would be in any way comparable to those of clinical internal medicine; in addition, they include a critique of Pinel's *méthode d'analyse*, which was limited to the minute observation of external signs (ibid., 133).[15] In contrast to this, Moreau presents a simulation of mental illness through hashish, which can invoke the analogy to the experience of sleeping and dreaming that is accessible to everyone. For though it may be said that dreams are lies, not even philosophers, particularly if they want to become sociologists, would entirely deny their existence.

In his book *La Psychologie Morbide dans ses Rapports avec la Philosophie de l'Histoire ou de l'Influence des Névropathies sur le Dynamisme intellectuel* (1859), he succinctly describes the procedure of his hashish experiments:

> Je suis *aliéniste*, et, de plus, m'étant volontairement plongé dans un état de folie *artificielle* (folie identique à la folie *spontanée*, du moins au point de vue des phénomènes psychologiques), j'ai pu me prendre moi-même comme sujet de mes observations. Alors pour moi la lumière s'est faite au sein des ténèbres; alors je me suis aperçu que toutes mes notions, laborieusement acquises depuis vingt-cinq années, sur la nature vraie, essentielle de la folie, étaient fausses et erronées.
>
> Il me fut démontré, dès lors, et ma conviction est la même aujourd'hui, que la folie n'était, en effet, [...] « que l'épanchement du songe dans la vie réelle;» [...] Je consacrait un long travail à la démonstration de cette vérité, au développement d'une idée qui m'était, pour ainsi dire, montée au cerveau avec les fumées enivrantes du hachisch. (Moreau de Tours 1859, 430n.)[16]

What connects these observations to a relevant corpus of data in the first place, however, is Moreau's assumption of a fundamental functional disorder or injury, a "lésion fonctionnelle primordiale" (Moreau de Tours 1845, 32), from which the different forms of madness are derived. He calls this disorder "fait primordial" (ibid., 31).

And this *fait primordial* is for him the essential and triggering factor, "le fait primitif et générateur de tous les autres" (ibid.). It consists, he says, in an over-excitation of intellectual abilities in their activity, or to put it more physiologically, in an oscillatory movement of nerve action (ibid., 99 f.).[17] This disposition, he continues, is as universal as possibly conceivable. According to Moreau, not only the sick person on the verge of madness, but also the slumbering person carried

---

[15] "Ici, comme dans beaucoup d'autres cas, l'observation a été superficielle, et partant, incomplète."

[16] With *l'épanchement du songe dans la vie réelle* Moreau quotes the poet Gérard de Nerval, who participated in the hashish experiments with Charles Baudelaire.

[17] "mouvement oscillatoire de l'action nerveuse".

through "the gates of horn and ivory"[18] as well as the test subject who feels the first effects of hashish all experience this condition. It is thereby no longer an enigma observed in awe from afar; it becomes, rather, a physiological condition with clearly describable features. Were the one held captive by madness to speak the language of the doctor after recovering clear consciousness, a scientific recording of this oscillatory movement of the nerves would be possible. This very possibility is due to hashish providing the doctor himself with access to an experience through which he can finally learn how to really understand madness in the first place: "[...] pour savoir comment déraisonne un fou, il faut avoir déraisonné soi-même; mais avoir déraisonné sans perdre la conscience de son délire, sans cesser de pouvoir juger les modifications psychiques survenues dans nos facultés" (ibid., 34).

In a novel interlacing of practices, which truly deserve the predicate modern, Moreau de Tours connects the domains of psychiatric anamnesis, of the "experience of the self," and of simulation, in that he lets the *observation intime*, the observation of the internal, run through the notational system of a *méthode d'analyse* that goes beyond Pinel's psychiatric analysis instead of through the filter of a philosophical analysis of the self. The doctor is no longer—or no longer absolutely—the one who, by virtue of his own faculty of reason, leads the patient back from mental alienation, but is rather the one who, now familiar with these confusions from his own experimental simulation, possesses the appropriate language to describe them:

> Par son mode d'action sur les facultés mentales, le hachisch laisse à celui qui se soumet à son étrange influence le pouvoir d'étudier sur lui-même les désordres moraux qui caractérisent la folie, ou du moins les principales modifications intellectuelles qui sont la point de départ de tous les genres d'aliénation mentale. (Ibid.)

Moreau's positing of a functional disorder as the cause of psychic illness is very revealing. He writes in *Du hachisch*:

> On voit, [...] que nous aussi, nous admettons une lésion fonctionnelle, non pas indépendante des organes, comme le croient les partisans de je ne sais quel *dynamisme* moral, mais liée essentiellement à une modification toute matérielle et moléculaire, quoique insaisissable de sa nature, insaisissable comme le sont, par-exemple, les changements qui surviennent dans l'intime texture d'une corde à laquelle on imprime des mouvements vibratoires d'intensité variable. (Ibid., 397)

True to this radical definition of mental illness as a *functional* disorder,[19] Moreau argues that while it is correct to assert that no functional disorder can appear without a pathological impairment of its organic basis, i.e., of the brain, as long as physiological-anatomical methods do not lead to a result, one must approach the problem from a different angle:

> Oui, incontestablement, des modifications (nous n'osons nous servir du terme de lésion) existent dans l'organe chargé des fonctions intellectuelles, mais ces modifications ne sont

---

[18] Good and bad dreams come to humans through these gates. Cf. Homer's *Odyssey* and Vergil's *Aeneid*.

[19] This general term refers to a pathological phenomenon without a causal anatomical-physiological correlate. In the cases treated by Moreau, one would today perhaps speak of functional psychoses or psychovegetative syndrome.

pas ce que l'on veut qu'elles soient généralement; et, sous la forme qu'on s'imagine et qu'on leur prête, elles échapperont toujours au recherches des investigateurs. Ce n'est pas dans telle ou telle disposition particulière, anormale, des diverses parties de l'organe de la pensée, disposition moléculaire, fixe, dont la texture de l'organe se trouverait altérée, qu'il faut les chercher, mais dans une altération de la sensibilité, c'est-à-dire l'action irrégulière, exaltée, diminuée, pervertie, de ces propriétés spéciales, d'où dépend l'accomplissement des fonctions intellectuelles. (Ibid., 397)

In these passages, Moreau walks a fine line between psychiatry and a newly emerging discipline in its own right, neurology. Accordingly, the argumentative separation between a pathological change in the brain that causes malfunction and a disorder in the intensity of this function itself (for which no direct anatomical-physiological correlate can be identified) appears difficult. Yet, for Moreau, it stands to reason that these pathological changes lie not in the topological structure of the brain but in a deeper level of the microstructure of the nervous system. The triumph of cellular pathology was, however, still to come: Rudolf Virchow's *Cellular Pathology in its Basis in Physiological and Pathological Histology/Die Cellularpathologie in ihrer Begründung auf physiologische und pathologische Gewebelehre* was not published until 1858. It was the first study to shift the focus to functionally specialized somatic cells as the smallest unit of anatomical pathology. Yet Moreau de Tours already strikes tones that would ring in the *Age of Nervousness*[20] in the second half of the century.

In any case, a test person under the influence of hashish undergoes neurological-psychic changes without losing his or her understanding forever; this person dreams without sleeping, perhaps becomes delirious, but can later remember the difference between these states and those of everyday existence; he or she personally experiences the fusion of dream and reality.

This self-observation illuminates an oscillation between "état de rêve" and "état hallucinatoire" (ibid., 351) that are transmitted through hearing since Moreau's drug of choice is hashish, which induces acoustic illusions and hallucinations in the majority of cases.[21] The poetic surge of the dream into the world of the real, cited by Moreau de Tours, is consequently regarded by him as a form of speech that makes audible for the one hallucinating what for the healthy person seems to be merely endless internal thinking: "Comme le rêveur, l'halluciné n'entendra pas seulement des sons qui auront autrefois frappé son oreille, mais il entendra des discours plus ou moins suivis. Dans l'état normal, penser c'est parler intérieurement; dans le cas où se trouve l'halluciné, c'est parler haut" (ibid., 352). Dreamer and hallucinator, according to Moreau, are inundated by speech, with a "thinking [...] clothed in the sensory sign of articulated sounds" (ibid., 354).[22]

When it comes to the sensual presence of the imagination, language, speech, and mental activity thereby imply a predominance of *fantasy* and *memory* in intoxication

---

[20] This being the title of an extensive study by Joachim Radkau (2000), who primarily investigates the phenomenon with a view towards the political and historical conditions in Germany.

[21] Moreau says himself in *Du Hachisch*, that the hearing is primarily affected by the hallucinations induced by hash (Moreau de Tours 1845, 180).

[22] "pensée, [...] revêtue du signe sensible des sons articulés".

as in madness (ibid., 63), whereas normality requires the suppression of these impulses: "[...] dissolution [...] désagrégation moléculaire de l'intelligence [...] décomposition intellectuelle" (ibid., 98) are thus clearly isolated diagnostic signs, which are not simply described as in the older *méthode de l'analyse,* but are thoroughly characterized through examining their expressions in word and text. With Moreau, nineteenth-century psychiatry takes a first revolutionary step towards regarding language, speech, and writing as media in the literal sense of the data of a dynamic psychology. Everyday thought, language in dreams, and acoustic hallucinations in states of intoxication or madness are comprehended from the physiological reality of their linguistic signs. Whereas "normal" life hides this reality under a coherent pattern that supports linearity and covers up or blocks out deviations, dreaming, intoxication, and madness do without these constraints and inner voices reemerge in speaking and writing (ibid., 66 ff.). Moreau illustrates this through the example of an afflicted person who hears voices:

> It is not enough that B. hears his thinking and internally speaks along with it, it is just as necessary that he performs the movements of the tongue and lips from which the articulation of the sounds results. One could truthfully say that nature is caught here in the very act. It is obvious how here the phenomenon of auditory hallucination is for him nothing other than enunciated thinking [...]. (Ibid., 354)[23]

The endless modifications of speech and of language, internal voices, thoughts, writing—all of that which for so long had formed and still forms the essence of the Western sense of self and its reproduction, is traced back by Moreau de Tours to its most intimate site and understood as an enunciation of physiological data from the brain and nervous system. This can only be provoked by a psycholytic drug through which the apparent unity of the *moi*, as predicted by Broussais, dissolves pharmacodynamically into its neurophysiological components. The preponderance of memory and imagination in hashish intoxication, which Moreau notes as a finding in his experiments (ibid., 63),[24] suggests that writing is in the end merely recording. What that also means for the doctor who is himself writing and recording his experiences under intoxication can only be guessed at.

What Moreau identified as specific for madness and dreaming, namely the interpenetration of the inner world of physiological perception and the physical outside world, worlds commonly averted to each other, can thus—based on the guidelines of language—be understood as signs of this interpenetration. Linguistic utterances thereby become objects of psychological medical science. In 1850, Moreau noted that while the insane write as they think,[25] individuals who are not insane but who produce insane writing are also of psychological relevance (Moreau de Tours 1850, 75). Even under the rule of a culture of the written word and with

---

[23] "Il ne suffit pas à B..., pour qu'il entende sa pensée, de la parler intérieurement, il faut encore qu'il exécute les mouvements de la langue et des lèvres d'où résulte l'articulation des sons. On peut dire véritablement qu'ici la nature est prise sur le fait. Il est évident que chez lui le phénomène de l'hallucination auditive n'est autre que la pensée parlée [...]."

[24] "[...] la mémoire et l'imagination prédominent [...]."

[25] 1850, p. 75: "[...] on peut dire que les aliénés *écrivent* comme ils pensent [...]."

their well-known features of preserving and forgetting, writing and speech remain practices that are susceptible to alteration, especially under the new primacy of psychological knowledge.

In such findings, which make writing and speech into a diagnostic medium of brain data, something uncanny dwells. This connection was not only evident in the field of psychiatry. The vast number of tremors, facilitations and collisions in the nineteenth-century metropolis, which are in fact not the hallucinations of a madman but the result of the overwrought nerves of a *décadent*, reveal a circuitry of nervous systems connected with other networks, for instance the transportation network. These networks "disperse" (*zerstreuen*) the self of the metropolis denizen in an utterly direct sense.[26]

In precisely this constellation, a young scientist in one of Europe's metropolises, drawing on studies of organic linguistic disorders, became interested in the dynamics of language and speech in terms of psychopathology. In *Studies on Hysteria/ Studien über Hysterie* (1895) Freud writes:

> I have not always been a psychotherapist. Like other neuropathologists, I was trained to employ local diagnoses and electro-prognosis, and it still strikes me myself as strange that the case histories I write should read like short stories and that, as one might say, they lack the serious stamp of science. (S.E. II, 160)

His dismay is perhaps a (self) forgetting of that which Freud was very well aware, namely, that novellas had begun to be read as clinical histories for quite some time in France. After all, Moreau de Tours' (1855) essay on the identity of dream and madness appeared in the bibliography of the first edition of *The Interpretation of Dreams* (S.E. V, 711). In a previous essay written in French, *L'hérédité et l'étiologie des névroses* (1896), Freud dealt with Moreau's still virulent thoughts regarding the heredity of neuroses, only accepting them, however, with reservations. His text *The Psychopathology of Everyday Life/Zur Psychopathologie des Alltagslebens* (1904) shows that the hypothesis of a parallel between dream and neurosis, as with that of a general nervousness, had remained especially relevant from a psychodynamic perspective. His French colleague already demonstrated this idea in 1850 in the slim volume *Un Chapitre Oublié de la Pathologie Mentale*.

References

Broussais, F. J. V. (1821). *Examen des doctrines médicales et des systèmes de nosologie: ouvrage dans lequel se trouve fondu l'examen de la doctrine médicale généralement adoptée, etc.; précédé de propositions renfermant la substance de la médicine physiologique* (Vol. 1–2). Paris: Méquignon-Marvis.

Broussais, F. J. V. (1828). *De l'irritation et de la folie, ouvrage dans lequel les rapports du physique et du moral sont établis sur les bases de la médecine physiologique*. Paris: Delaunay.

Cabanis, P.-J.-G. (1798). *Du degré de certitude de la médicine*. Paris: Crapelet.

---

[26] The etymology of "zerstreuen, Zerstreutheit" does in fact have psychopathological roots. (Cf. Röhrich 2004, 1203 ff.).

Cabanis, P.-J.-G. (1799). *Traité du physique et du moral de l'homme* (Vol. 1). Paris: l'Institut national.
Comte, A. (1864). *Cours de philosophie positive* (2nd ed.). É. Littré (Ed.). Paris: J.-B. Baillière et fils. (1st edition published in 1830.).
Cullen, W. (1785). *Synopsis nosologiae methodicae* (Vol. 1). Edinburgh: Prostant venales apud Gulielmum Creech.
Ellenberger, H. F. (1970). *The discovery of the unconscious: The history and evolution of dynamic psychiatry.* New York: Basic Books.
Esquirol, J. É. D. (1838). *De maladies mentales considérées sous les rapports médical, hygiénique et médico-légal* (Vol. 2). Paris: J.-B. Baillière.
Foucault, M. (1973). *The birth of the clinic: An archaeology of medical perception* (A. M. Sheridan, Trans.). London: Tavistock. (Original work published in French 1963)
Foucault, M. (2006). *Madness and insanity: History of madness in the classical age* (J. Khalfa, Ed., Trans. & J Murphy, Trans.). New York: Routledge. (Original work published in French 1961)
Freud, S. (1891a). *Klinische Studie über die halbseitige Cerebrallähmung der Kinder.* Wien: Moritz Perles.
Freud, S. (1891b). *Zur Auffassung der Aphasien: Eine kritische Studie.* Leipzig: Franz Deuticke.
Freud, S. (1895). *Studies on hysteria.* S.E II.
Freud, S. (1896). L'hérédité et l'étiologie des névroses. *Revue Neurologique, 4*(6), 161–169.
Freud, S. (1900). *The interpretation of dreams.* S.E. V.
Freud, S. (1904). *The psychopathology of everyday life.* S.E. VI.
Gall, F. J. (1809). *Untersuchungen über die Anatomie des Nervensystems überhaupt, und des Gehirns insbesondere. Ein dem französischen Institute überreichtes Mémoire.* Paris/Strasburg: Treuttel & Würtz.
Gall, F. J. (1822). *Sur les fonctions du cerveau et sur celles de chacun de ses parties.* In Erna Lesky (Ed., 1979), *Franz Joseph Gall: 1758–1828. Naturforscher und Anthropologe.* Bern/Stuttgart: H. Huber.
Lechler, W. H. (1960). *Neue Ergebnisse in der Forschung über Philippe Pinel. Seine Familie, seine Jugend- und Studienjahre 1745–1778* (Doctoral dissertation). Munich: Self-published.
Lesky, E. (Ed.). (1979). *Franz Joseph Gall: 1758–1828. Naturfor-scher und Anthropologe.* Bern/Stuttgart: H. Huber.
Magendie, F. (1841). In C. James (Ed.), *Leçons sur les fonctions et les maladies du système nerveux* (Vol. 2). Paris: Le Caplain.
Moreau de Tours, J.-J. (1845). *Du hachisch et de l'aliénation mentale: Études psychologiques.* Paris: Fortin Masson.
Moreau de Tours, J.-J. (1850). *Un Chapitre Oublié de la Pathologie Mentale.* Paris: Masson.
Moreau de Tours, J.-J. (1855). De l'identité de l'état de rêve et de la folie. *Annales médico-psychologiques, 3e série,* (I), 1, 361–408.
Moreau de Tours, J.-J. (1859). *La Psychologie Morbide dans ses rapports avec la philosophie de l'histoire ou de l'influence des névropathies sur le dynamisme intellectuel.* Paris: Masson.
Müller, J. (1822). *Nemo psychologus nisi physiologus.* Bonn.
Pinel, P. (1798). *Nosographie philosophique ou la méthode de l'analyse appliquée a la medicine.* Paris: Maradan.
Pinel, P. (1801). *Traité médico-philosophique sur l'aliénation mentale ou la manie.* Paris: Brosson.
Pinel, P. (1802). *La médecine rendue plus précise et plus exacte par l'application de l'analyse.* Paris: Brosson.
Popp, W. (1981). *Wege des exakten Denkens.* Munich: Ehrenwirth.
Radkau, J. (2000). *Das Zeitalter der Nervosität: Deutschland zwischen Bismarck und Hitler.* Munich: Econ-Ullstein-List-Verl.
Röhrich, L. (2004). *Lexikon der sprichwörtlichen Redensarten* (Vol. 2). Darmstadt: Wissenschaftliche Buchgesellschaft.
Scharbert, G. (2010). *Dichterwahn: Über die Pathologisierung von Modernität.* Munich: Fink.
Virchow, R. (1858). *Die Cellularpathologie in ihrer Begründung auf physiologische und pathologische Gewebelehre.* Berlin: August Hirschwald.

Chapter 6
Dreams, Unconscious Fantasies and Epigenetics

Tamara Fischmann

Abstract Fischmann presents an interdisciplinary combination of psychoanalysis and neuroscience, in which she focuses on different approaches towards dreams, the dreaming mind, and the brain from a psychoanalytical, neuropsychoanalytical and neurobiological stance. The current FRED study continues to investigate changes in brain functions in chronically depressed patients after long-term therapies, looking for multi-modal neurobiological changes in the course of psychotherapy. Data from both neurobiology and psychoanalysis suggest that emotionally meaningful life experiences are encoded in memory by sensory percepts. These encoded memories will then recur in dreams. Therefore, dreaming can no longer be considered as random and meaningless. The author further links dreams and unconscious fantasies with epigenetics. The fact that epigenetic regulation, that is, chromatin remodeling in neurons, not only occurs in the developing brain but also in the mature, fully differentiated brain, raises questions about psychodynamic interactions in the developing mind that we are just now beginning to understand.

Keywords Dreams • FRED study • Epigenetics • Mind-brain relationship • *Nachträglichkeit* • Nodal images • REM sleep • "Dream of the botanical monograph"

What do dreams, unconscious fantasies, and epigenetics have in common? Epistemologically speaking, dreams and unconscious fantasies belong to the theoretical domain of psychoanalysis, while epigenetics belongs to the field of neuroscience as part of biology. All three deal, in one way or another, with the mind (soul) and brain and probe the different functions of each. But how and by what means is this done? The article gives some insights into an ongoing research project where psychoanalysis *meets* neuroscience. The first section focuses on the topic of dreams and elaborates the different approaches towards dreams and the dreaming mind and brain from a psychoanalytical, neuro-psychoanalytical and neurobiological stance. In order to illustrate the fruitfulness of the interdisciplinary combination

---

T. Fischmann (✉)
Sigmund-Freud-Institut, Beethovenplatz 1-3, D-60325 Frankfurt am Main, Germany
e-mail: dr.fischmann@sigmund-freud-institut.de

of the neurosciences with psychoanalysis, the second section gives a brief look at a current research project. In the final two sections the relationship between dreams and unconscious fantasies and dreams and epigenetics is explained.

6.1 Dreams

In 1895 Sigmund Freud set out to understand the mind-brain relationship in his *Project for a Scientific Psychology/Entwurf einer Psychologie*. He was very much guided by his neurological education but he was also striving to understand how psychic processes, consciousness, mental processes and neurons work together and tried to establish his findings scientifically.

He proposed that a system he called *Phi* (φ) is responsible for perception. The *Psi* (ψ) system, he claimed, contains memory and endogenous stimuli, while the *Omega* (ω) system represents reality. He further designated systems for motility (M) and for perception-consciousness (P-C). He described the dynamics of these systems in the following terms: An outside stimulus reaches perception via φ from where it moves to the motility system M while being deposited in memory and integrated into the world of endogenous stimuli to create the unconscious psychic system ψ. This process holds true, so he claimed, for the waking state, where reality ω guides the progress along this path. As for the sleeping state, where movements are more or less inhibited, psychic energy will build up in the psychic system ψ, which cannot proceed towards M and is forced to egress via the gateway of perception φ. This perception, however, has no external object, a phenomenon which Freud conceived of as a hallucination meant to satisfy a repressed desire, i.e., it relieves the built up psychic energy in ψ (Fig. 6.1).

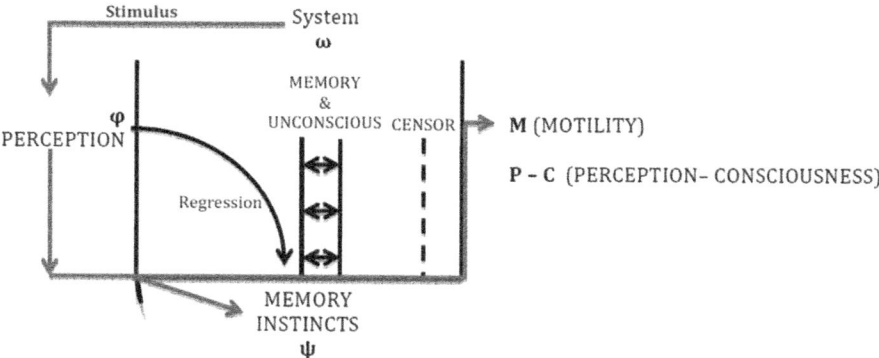

Fig. 6.1 Model of mind-brain relationship proposed by Freud in the *Project for a Scientific Psychology* (1895) and in *The Interpretation of Dreams* (1900)

On the basis of this model Freud formulated his definition of the dream in the *Interpretation of Dreams/Die Traumdeutung* (1900) as a hallucinatory satisfaction of a desire repressed in infancy. Consequently, dreams enable the unconscious to proceed unnoticed by the censor to the P-C system according to Freud. In order for dreams to become conscious they have to undergo distortions (*Entstellungen*) and transformations imposed by censorship (condensation, displacement, symbolization, dramatization just to name a few). Therefore, we might assume that censorship not only creates dreams but also makes us forget them. Gaining access to dreams is usually accomplished by narrating them, i.e., by transferring dream images into a narrative language. Via an associative syntagmatic chain the dream generates a linguistic "swap-over" that allows one proposition to take the place of another. Freud conceptualized the latter as a primary representation transformed through secondary processing. He also wrote that dream-work is permeated with affectivity, which enables an emotional recovery of memories from early childhood and called this process *Nachträglichkeit*. Its aim is to transcribe the history of traumatic events from a patient's childhood as a "rewriting of memory". It thus embodies the aim of dream-work itself.

Dream theory has evolved since Freud. Melanie Klein's work (1948) brought about a shift that supplanted Freud's instinctual energetic model based on desire and its repression with a relational model based on more complex modalities in the organization of personality and the mind's unconscious functions. Wilfred Bion (1963) claimed dreams were a tool of the mind, such that mental functions continue to be active in waking states where they are dominated by fantasies, while during sleep states they are dominated by dreams. From this angle, dreams were seen as a way of putting "on-stage" the affects involved in the relation of the here and now. Today dreaming, understood as a mechanism for rewriting memory, might be considered the most credible and reliable tool in reassigning significance to a past experience, even if that experience was pre-verbal and pre-symbolic.

6.1.1 Dreams in Psychoanalysis and Neurobiology

In psychoanalysis dreams deliver data about the dreaming mind; in neurobiology they give information about the dreaming brain. Both are important if one wants to understand the relationship between the mind and the brain. The concepts from the science of mental and psychic states and brain sciences employ different vocabularies and techniques. They are framed at different levels of abstraction, and their units are not interchangeable. Experimental data gathered in both fields reveal covariance and correspondences rather than cause-and-effect sequences when set in relation to one another. Thus, for instance, the rich neurobiological data on REM-sleep in animals provide a wealth of information about the dreaming *brain* but nothing about the dreaming *mind*. So, one task might be to explore the relationship between the two different data sets, for example, with a *bottom up* approach regarding neurobiological data and a *top down* approach for the psychoanalytic and psychological data. The following sections clarify this relationship between neurobiological data on the dreaming brain and psychoanalytic data on the dreaming mind.

6.1.2 The Dreaming Mind and Brain

John Allan Hobson and Robert W. McCarley (1977) assert that Freud's theory of dreams as being instigated by a wish has been refuted. They propose instead an "activation-synthesis" hypothesis of dreaming, in which they claim that ascending excitatory ponto-geniculo-occipital waves stimulate higher midbrain and forebrain cortical centers producing rapid eye movements. The resulting activation that randomly spreads over the association cortex then activates stored traces of memory. This process constitutes the activation part of their theory, which further claims that dream images and experiences are randomly generated and do not convey any meaning. To make sense of the dreaming experience, the dream "edits" the story line to make sense of it upon waking. This editing process constitutes the synthesis part of their hypothesis. With respect to affects in a dream they initially regarded them as a secondary response to content, but in a later paper Hobson (1999) modified this view in light of new imaging data and claims that affects in a dream are primary in shaping its content.

Philosophically speaking there is a logical problem here concerning the way the neurobiological data has been interpreted. It is true that this data demonstrates an altered balance in the relative levels of aminergic and cholinergic activity in the pons, which triggers the onset of REM sleep. But it is not true that this altered balance instigates the dream per se as a mental experience. Hence, it cannot disprove Freud's hypothesis that a wish is the instigator of a dream.

To take this argument of dream instigation a step further, data from both neurobiology and psychoanalysis strongly suggest that emotionally meaningful life experiences are encoded in memory by sensory percepts that were registered during the life experience that they encode. Mortimer Mishkin and Tim Appenzeller (1987) were able to show a link between affects and memory processing. That is to say, a specific pattern will not be available for matching in a declarative memory until it has been linked by limbic and paralimbic structures to the memory circuits of the prefrontal and association cortices. It is important to note here that only by passing through the limbic structures that generate and regulate emotion can the encoded perceptual trace become inextricably linked to the affect that accompanied its processing into memory. This finding coincides with the idea of enduring neural networks of memory organized by emotions (Reiser 1984, 1990).

These kinds of networks are organized around a core of perceptual images that encode memories of early events that were experienced as highly emotional, some even catastrophic, by the child. As development proceeds, the networks branch out as a result of later events with similar conflicts and emotional states. The corresponding encoded images are closely interconnected and strongly related to other images in the network and through them to others still. This relation can be thought of as nodal points in the enduring memory networks of the mind and brain. From this perspective it becomes obvious that memory is organized by emotion, and this is closely related to dreaming as I argue in the following.

When examining the data on the dreaming brain elicited by PET-studies, one finds that the same limbic and paralimbic areas involved in the matching function that links affect and memory (cf. Mishkin and Appenzeller 1987) are highly activated during REM sleep (Braun et al. 1998; Maquet et al. 1996; Nofzinger et al. 1997). Thus, the corticolimbic circuit linking percept to affect described by Mishkin's experiments may be considered to constitute an "off-line processing" function in REM sleep, as Edmond M. Dewan (1970) and Jonathan Winson (1985) have shown. This "off-line process" serves a crucial evolutionary function in terms of adaptation and survival by making it possible to store significant information during sleep and discard that which is insignificant. But far more important here is that the process indicates that these areas must be major participants in generating affect in dreams and thus attests to the notion that dreams take part in memory formation, or rather memory consolidation, via affects.

This notion finds support in the idea of enduring neural networks of memories in the brain (cf. Reiser 1990), which seems to be congruent with the psychoanalytic concept of enduring nodal memory networks in the mind. Detailed examples of memory networks described by patients give rise to the postulation that dream images may be recruited from memory by affects that connect them to a current conflict. In other words, the images might have been sensitized to ponto-geniculo-occipital wave stimulation by the relevant affect rather than appearing in the dream as a result of random stimulation during REM sleep.

To give an example of this idea of nodal images, let us consider Freud's "Dream of the botanical monograph" (1900). The ideas and memories represented in this dream seem to be arranged around such nodal networks. His ideas produced through free associations with the dream elements connect to either the nodal point "botanical" or "monograph" which in turn lead him back through a network of memories of failures that are accompanied by emotions of shame and guilt. These affects can be regarded as organizers of the memory network, spinning associative threads as connective links between images that were encoded with earlier and previously painful memories. From here it seems obvious that a dream is more than an "electric brain storm" imbued with meaning only after the fact.

Many clinical psycho-physiologic experiments (Palombo 1973, 1978; Greenberg and Pearlman 1974; Greenberg et al. 1990; Cartwright 1977, 1991) proved to be highly productive in a different way: Their findings reveal correspondences between qualitative and cognitive aspects of dream content and the emotional problems that confronted the dreamer in broad time frames that encompass stressful life problems from the past and the present. These correspondences further emphasize the role of memory and emotion in the dream process. Given that emotion occupies both domains (mind and brain), it may provide the key to understanding the mechanisms that link covariant as well as corroborating data on the mind and the brain.

Mark Solms (2000), in his extensive clinical anatomy research on the neuropathology of clinical sleep and dreaming disturbances, noted that Freud's definition of "wish" characterizes the term as a psychobiologic phenomenon; a wish is the expression of an endogenously stimulated need with a somatic origin (Freud 1900, Stud. II, 565). Thus, in designating the wish as the instigator of a dream, Freud did not think of the wish as an immaterial mental phenomenon but rather as a source of motivation in terms of material or energy, which arises from within the body as part of the metabolic life process and creates a bodily need. This need, then, requires work on the part of the mental apparatus (the mind) to satisfy it. This is supported by at least two studies regarding dream cessation. Solms was able to show that dreams can also occur in the absence of REM sleep and that REM sleep can occur in the absence of dreaming. He also showed that, in fact, only two forebrain lesion sites are associated with a cessation of dreaming. In addition, Jaak Panksepp (1998) identified a part of the brain (the bilateral ventromesial quadrant of the frontal lobes) as the "curiosity-interest-expectancy" command system, a system associated with endogenously stimulated, appetitive craving states. This part of the brain is also an area of greater activity during REM sleep. Leucotomy of this area results in anhedonia, a lack of initiative and a cessation of dreaming, even though REM sleep continues. Thus, this part of the brain and its function are part of the dreaming brain. The correlation of dreams with states of curiosity and interest found here must be acted upon during sleep if one does not want to wake up. Producing a dream seems to be the best solution for this, and therefore dreaming can be categorized as a significant mental act.

6.2 When Psychoanalysis Meets Neurophysiology

The ongoing Frankfurt fMRI/EEG Depression study (FRED)[1] illustrates the potential for a productive combination of the two domains of psychoanalysis and neuroscience. This very ambitious project is currently being conducted at the *Sigmund-Freud-Institute* (SFI) and the Brain Imaging Center (BIC) in cooperation with the *Max-Planck-Institute for brain research* (MPIH)[2] in Frankfurt. The project investigates changes in brain functions in chronically depressed patients after long-term therapies with the goal of finding multi-modal neurobiological changes over the course of psychotherapy.

When looking at depression from a physiological angle, some interesting findings have been put forth. For instance, several studies show that depression is related to a neurotransmitter disorder or a frontal lobe dysfunction (cf. Belmaker and Agam 2008; Caspi et al. 2003; Risch et al. 2009). Northoff and Hayes (2011) have convincingly suggested that the so-called "reward system" is disturbed by depression and that there is evidence that deep brain stimulation can improve severe depression. But despite all these findings no distinct physiological marker for depression has been found in the brain. Accordingly, it seems plausible to pose a research question about whether changes in the course of therapy have physiological correlates in the brain, a question that we are currently investigating at FRED.

---

[1] Funded by the Neuro-Psychoanalysis Society—HOPE (M. Solms, J. Panksepp et al.) and the Research Advisory Board of the IPA.
[2] We are grateful to the BIC and MPIH (W. Singer, A. Stirn, M. Russ) and the Hanse-Neuro-Psychoanalysis-Study (A. Buchheim, H. Kächele, G. Roth, M. Cierpka et al.) and LAC—Depression Study for supporting us in an outstanding way.

Generally speaking, psychotherapists, especially psychoanalysts, work with what patients can remember and with recurring and usually dysfunctional behaviors and experiences. We assume that this has consequences within the brain, like synapse configuration, priming, axonal budding and others. This led us to hypothesize that (1) psychotherapy is a process of change in the encoding conditions of memory, and (2) elements of memories can be depicted in fMRI by a recognition experiment involving memories related to an underlying conflict.

In order to investigate this, we recruited chronically depressed patients and conducted two interviews with them in the first diagnostic phase of the study. The first was an Operationalized-Psychodynamic-Diagnostics (OPD) interview that concentrates on axis II (relational), and the second was a dream interview (see Fig. 6.2). Based on these two interviews, we then designed the stimuli for the fMRI-scanning individually for each patient. To do this we began by taking dream-words from a significant dream in the dream interview and then we took confrontational sentences as formulated in the OPD interview. Brain activation patterns resulting from these stimuli in the fMRI serve as dependent variables (DV). Measurements were taken at three different time points revealing changes in activation patterns that occur over the course of therapy (see Fig. 6.3).

Currently, we have recruited ten chronically depressed patients (from the PA-group), from whom we have obtained 17 dreams so far. If we assume that a dream is a linguistically coded memory, it will contain some conflict-laden material if it is a significant one. It will also have portentous emotional qualities that are related to the primary dream process, and these dream-words have a different quality than words taken from an "all-purpose" story. In the fMRI, when those *dream-words* are recognized, they reactivate dream-specific encoding conditions in certain brain areas. This memory recognition task with dream-words is the dream experiment part of FRED.

In fact, the results of our dream experiments reveal that patients confronted with dream-words (in contrast to so-called neutral words taken from an all-purpose story) showed differential activation of the precuneus, the ventro-lateral pre-frontal cortex (VLPF) and the anterior cingulate gyrus among others. These three brain areas are involved in self-processing operations (e.g., experiences of self-agency), generating basic causal explanations and regulating emotions.

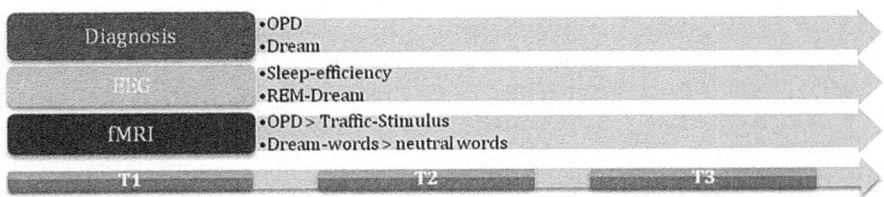

Fig. 6.2 Time schedule and measurement points within FRED

Factor A: Type of Therapy	Patient	Patient	Control
DV:	PA	CBT	---
sleep-efficiency	a_1	a_2	a_3
REM-dream	b_1	b_2	b_3
dream-words	c_1	c_2	c_3
interaction-conflict (OPD)	d_1	d_2	d_3

Fig. 6.3 FRED design—1-factorial design with repeated measurement and control group. *PA* Psychoanalytic treatment-group, *CBT* Cognitive-behavioral treatment-group

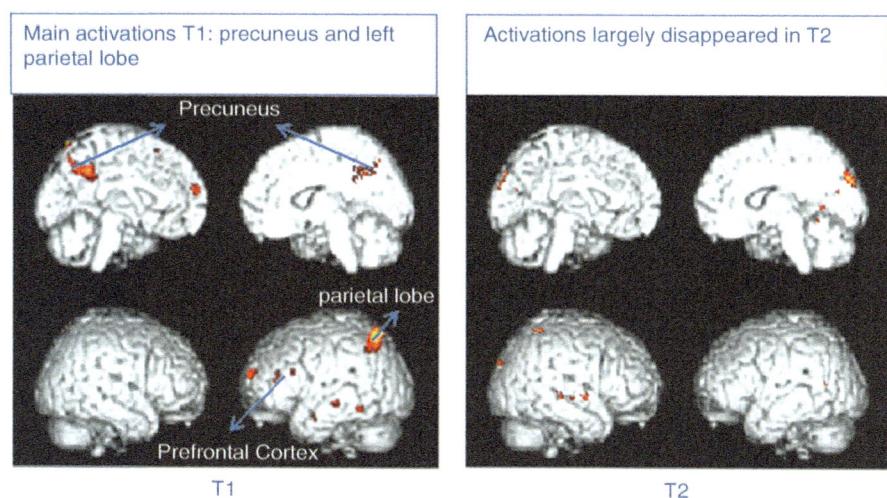

Fig. 6.4 T1 and T2 comparison of dream-word recognition in a single case

Over the course of therapy we were able to show that that recognition or rather re-sounding of initially significant dream content at the beginning of therapy specifically activated the precuneus and the left parietal lobe. This was not the case after one year of therapy. The disappearance of these activation areas at T2, areas which are significant to emotional processing by the self, support the hypothesis that the dream content has possibly lost its special importance and is experienced now in the same manner as the neutral story (see Fig. 6.4).

The OPD part of FRED consists of three conditions in the fMRI-scanner, and each is repeated six times. In condition 1, four subjectively confrontational (conflict-oriented) statements extracted from a previous OPD interview (relational axis II) are presented consecutively on a screen in the fMRI-scanner. In condition 2, subjects see four statements from an all-purpose situation presented in the same manner; and finally, condition 3 is composed of four relaxation statements (see Fig. 6.5).

6 Dreams, Unconscious Fantasies and Epigenetics

	Condition 1	Condition 2	Condition 3
1.	Most of the time I had to control myself and to manage by myself	A traffic participant is acting wrong	Think of a safe place
2.	Now I feel very lonesome and someone to take of me	You are annoyed with him	Relax
3.	I can bear closeness only badly and	You react	Get your head free
4.	don't think that anyone is really interested in me	He reacts inappropriately	Don't think about anything

Fig. 6.5 The three conditions of stimulus presentation in the fMRI

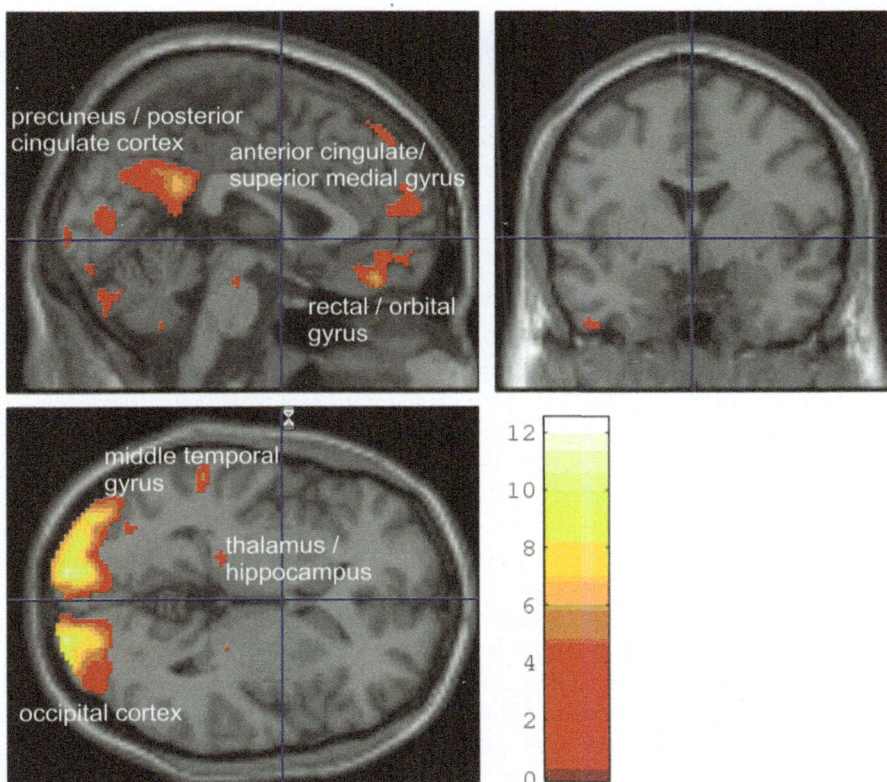

Fig. 6.6 fMRI-Scans contrasting dysfunctional sentences (condition 1) > traffic (condition 2) + relaxation (condition 3). Second level analysis p < .05, FDR corrected; N = 13

By contrasting the different conditions (dysfunctional sentences > traffic + relaxation), the analysis of the fMRI brain scans revealed specific activation patterns in the precuneus, posterior and anterior cingulate gyrus, medial frontal cortex (MFC), occipital cortex and the left hippocampus for condition 1 (dysfunctional sentences) (see Fig. 6.6). The occipital cortex and precuneus are important brain structures for primary visual processes (occipital c.) and visual-spatial imagery (precuneus).

Apart from this, the precuneus is also known to be an important area of the brain for episodic memory retrieval and self-processing operations, i.e., for assuming a first-person perspective and for experiencing agency. The cingulate gyrus, an important part of the limbic system, helps regulate emotions and pain and constitutes an important feature of memory, just like the hippocampus that is aligned with memory formation, specifically long-term memories. The MFC is postulated to serve as an online detector of information processing conflicts (Botvinick et al. 2004) but it also has a regulative control function for affective signals (Critchley 2003; Matsumoto et al. 2003; Posner and DiGirolamo 1998; Roelofs et al. 2006; Stuphorn and Schall 2006). In a single case study we were also able to show that MFC activation could no longer be detected after one year of psychotherapy, which suggests that the conflict impact has diminished in the course of therapy.

In summary, data from both neurobiology and psychoanalysis strongly suggests that emotionally meaningful life experiences are encoded in memory by sensory percepts that were registered during the life experience that they encoded. These encoded memories will recur in dreams and tap into memories in the same area of brain. Therefore, dreams can no longer be considered to be randomly generated and to not convey any meaning.

6.3 Dreams and Unconscious Fantasies

Another field closely related to dreams is that of unconscious fantasies. Unconscious fantasies have long been associated with the notion of a *dynamic unconscious*. Both are only accessible via their derivatives and require inference from manifest evidence just like in dreams, in which one is able to differentiate between the manifest and latent content of a dream. For Freud, fantasies were *motivated* by drives and *structured like dreams* by *primary process*. They also *function* to maintain internal affective equilibrium and to help adapt to outer reality. Just like with dreaming, the concept of *unconscious fantasies* has changed along with the evolution of psychoanalysis in its diverse directions since Freud. Specifically, its regulation by primary process thinking, which links unconscious fantasies to early childhood, has led to much debate from within psychoanalysis but also from developmental psychology and from neuroscientific findings on memory. Findings in developmental psychology and neuroscience regarding memory shed light specifically on primary process thinking and give way to a deeper understanding of dream processes. Among other features, primary process thinking is regulated by *condensation* and *timelessness*, features we often encounter in dreams. If the claim that unconscious fantasies derive from early childhood experiences holds true, an appropriate question might be posed as to whether the assumption about how unconscious fantasies relate to primary process thinking can be correctly deduced from the way young children think. When looking at concept formation in childhood and condensation as encountered in adult dreaming, we find major differences. Indeed, condensation is considered to stem from "some mechanical or economic factor" of dreaming itself, as Freud puts it in his *Introductory Lectures on Psychoanalysis/Vorlesungen zur Einführung in die Psychoanalyse* (1916, S.E. XV, 173) and is not caused by dream censorship, i.e., it is due instead to constraints of visual representation. A good example for this is again Freud's dream of the *botanical monograph* in the *Interpretation of Dreams/Die Traumdeutung* (Freud 1900): "I had written a monograph on a certain plant. The book lay before me and I was at the moment turning over a folded coloured plate. Bound up in each copy there was a dried specimen of the plant, as though it had been taken from an herbarium" (S.E. IV, 169). If we scrutinize one element more closely, namely *monograph*, we can see that it is a temporary category. Freud associates it with medical books, specifically to monographs on the anesthetic properties of cocaine or to the *Festschrift*. The element monograph is thus by no means a perceptual category as its associations all refer to something *written* (-graph, -schrift) by or about or for one (mono-) subject or object. In young children's conceptual thinking, we would not expect to find this kind of temporary category based on grammatical permutations, which they are still trying to master at this age. Thus, condensation may be useful for interpreting temporary categorizations in the dreams of adults. But condensation does not illustrate the processes of early childhood categorization or concept formation. The latter, we might assume, stems from the way Freud came across condensation in dreams, that is, by looking at adult dreams and not by observing young children's concept formation.

Another feature of unconscious fantasies is its *timelessness*. One might be justified in asking whether this feature also holds true for children's mentation. From developmental studies we know that during a child's second year it becomes able to express itself verbally and to understand both time and duration. And as Jean Piaget (1924) has indicated, the ability to verbally express these two components of time is based on cognitive knowledge established as early as the sensorimotor period during the first two years. Thus, a sense of time already begins to be constructed during infancy, and we might conclude that children do indeed have a sense of time.

Timelessness, as we psychoanalysts know it with its (missing) features of tense and aspect, may be enlisted defensively by a subject, as has been shown with respect to *trauma* (cf. Bonomi 2003; Laub 2005). One reaction to a trauma may be to *project it forward as a fear of a future occurrence,* thereby undoing/denying the past trauma and making its mastery possible. Unfortunately, the trauma must be kept present in order to serve as a signal of a future danger. This aspect of timelessness also applies to children's mentation insofar as the persistency of unconscious fantasies is understood as related to "long-forgotten wishes of childhood" in "their pristine form, vital and unchanged" (Arlow and Brenner 1964). In terms of thinking and speaking, unconscious fantasies persist because they are unmarked by tense and aspect. As they are not accessible to communication, they exist as propositions without performative force. Moreover, unconscious fantasies represent children's theories about their reality. Their theories are based on their own feelings and actions, actual and distorted perceptions, the perspectives of others with whom they identify, and the language and cultural practices they incorporate. What cannot be interconnected, and thereby not be given meaning, remains alien and potentially overwhelming and frightening. Possible solutions include *splitting off* what cannot be understood, *projecting* it onto another person, *denying* its existence or *disavowing* its meaning. Unconscious fantasies thus contain both the child's naive view of the world, some of which is itself formulated defensively, as well as later defensive revisions of this view.

Highlighting the dynamics of unconscious fantasies from the perspective of the brain's neuroplasticity and epigenetics is a challenging but promising endeavor still in an embryonic state.

6.4 Epigenetics

Epigenetics largely deals with the question of how nature and nurture interact. This question has been debated in psychology for over two centuries (Collins et al. 2000) in the effort to discern the relative importance of an individual's innate qualities versus the role of personal experience in determining individual differences in behavioural traits. Specifically, *behavioural epigenetics* applies principles of epigenetics to the study of physiological, genetic, environmental and developmental mechanisms in human behaviour (Lester et al. 2011). Epigenetic regulation can be generally regarded as chromatin remodelling in neurons, in which the activity of a particular gene is controlled by the structure of chromatin in proximity to that gene. The process is crucial for nervous system development. Epigenetic regulation does not only occur in the developing brain but also in the mature, fully differentiated brain; thus, this field of research promises to improve the understanding of brain plasticity.

Psychodynamic models emphasize how children are adversely affected if their relational needs are not satisfied and show that socialization has enormous effects on development. What role the approximately 20,000–25,000 genes of the human genome play here is now beginning to be better understood through epigenetic research. Although our genes deliver the genetic blueprint, determining influences on gene activity come from external factors, i.e., molecular regulating mechanisms that are not part of the DNA within but rather attached to it. These epigenetic factors control genomic functions by turning genes on or off, depending on, for instance, early childhood experiences of maltreatment. Patrick McGowan et al. (2009) showed that a key-gene in the hippocampal cells of suicide-victims who suffered childhood abuse no longer functioned adequately. The gene itself was not damaged but was turned off via a chemical marker. McGowan and Moshe Szyf (2010) suggest that epigenetic processes might mediate the effects of the social environment during childhood on hippocampal gene expression, but they also caution that, "It remains unclear whether the epigenetic aberrations documented in brain pathologies were present in the germ line, whether they were introduced during embryogenesis, or whether they were truly changes occurring during early childhood" (70).

Another question of interest here was raised by Natan P.F. Kellermann (2011) who asks whether nightmares can be inherited? In studying the transgenerational transmission of trauma he points out that although bad dreams originate from many factors, nightmares in children that resemble parental trauma suggest the presence of epigenetic transmission—but only as a general tendency for having frightening nightmares. The content of a specific nightmare might be affected by epigenetic marks, but its determinants may still be rooted in the individual's experiences that left these epigenetic marks in the genome. This assumption still needs further investigation.

To conclude, psychic determinism, i.e., the discovery that human actions might be more completely understood by ascribing unconscious wishes and beliefs to conscious states, is considered by many to be Freud's greatest achievement (cf. Hopkins 1992; Wollheim 1999). And as Eric Kandel (1999) has pointed out, the interdisciplinary dialogue between the neurosciences and psychoanalysis may well serve to determine the validity of principles from the psychoanalytic theory of the mind as established by Freud in the early twentieth century. This will allow us to reformulate psychoanalytic concepts in light of emerging biological data as Freud himself recommended in *Beyond the Pleasure Principle/Jenseits des Lustprinzips* (1920): "The deficiencies in our descriptions would probably vanish if we were already in a position to replace the psychological terms by physiological or chemical ones" (S.E. XVIII, 60). For now this endeavour is still coming of age, and we are only just beginning to understand the epigenetic and psychodynamic interactions in the developing mind.

References

Arlow, J. A., & Brenner, C. (1964). *Psychoanalytic concepts and the structural theory*. Oxford: International Universities Press.
Belmaker, R. H., & Agam, G. (2008). Major depressive disorder. *New England Journal of Medicine, 358*(1), 55–68.
Bion, W. (1963). *Elements of psycho-analysis*. London: Heinemann.
Bonomi, C. (2003). Between symbol and antisymbol – The meaning of trauma reconsidered. *International Forum of Psychoanalysis, 12*(1), 17–21.
Botvinick, M. M., Cohen, J. D., & Carter, C. S. (2004). Conflict monitoring and anterior cingulate cortex: An update. *Trends in Cognitive Sciences, 8*(12), 539–546.
Braun, A. R., Balkin, T. J., Wesensten, N. J., Gwadry, F., Carson, R. E., Varga, M., Baldwin, P., Belenky, G., & Herscovitch, P. (1998). Dissociated patterns of activity in visual cortices and their projections during human rapid eye movement sleep. *Science, 279*, 91–95.
Cartwright, R. D. (1977). *Night life: Explorations in dreaming*. Oxford: Prentice-Hall.
Cartwright, R. D. (1991). Dreams that work: The relation of dream incorporation to adaptation to stressful events. *Dreaming, 1*(1), 3–9.
Caspi, A., Sugden, K., Moffitt, T. E., Taylor, A., Craig, I. W., Harrington, H. L., McClay, J., Mill, J., Martin, J., Braithwaite, A., & Poulton, R. (2003). Influence of life stress on depression: Moderation by a polymorphism in the 5-HTT gene. *Science, 301*(5631), 386–389.
Collins, W. A., Maccoby, E. E., Steinberg, L., Hetherington, E. M., & Bornstein, M. H. (2000). Contemporary research on parenting: The case for nature and nurture. *American Psychologist, 55*(2), 218–232.
Critchley, H. (2003). Emotion and its disorders. *British Medical Bulletin, 65*, 33–47.

Dewan, E. M. (1970). The programing (P) hypothesis for REM sleep. *International Psychiatry Clinics, 7*(2), 295–307.
Freud, S. (1895). *Project for a scientific psychology*. S.E. I.
Freud, S. (1900). *Interpretation of dreams*. S.E. IV.
Freud, S. (1916). *Introductory lectures on psychoanalysis*. S.E. XV.
Freud, S. (1920). *Beyond the pleasure principle*. S.E. XVIII.
Greenberg, R., & Pearlman, C. (1974). REM sleep and the analytic process: A psychophysiologic bridge. *Psychoanalytic Quarterly, 44*, 392–402.
Greenberg, R., Pearlman, C., Blacher, R., Katz, H., Sashin, J., & Gottlieb, P. (1990). Depression: Variability of intrapsychic and sleep parameters. *The Journal of the American Academy of Psychoanalysis, 18*, 233–246.
Hobson, J. A. (1999). The new neuropsychology of sleep: Implications for psychoanalysis. *Neuropsychoanalysis, 1*, 157–183.
Hobson, J. A., & McCarley, R. W. (1977). The brain as a dream state generator: An activation-synthesis hypothesis of the dream process. *American Journal of Psychiatry, 134*(12), 1335–1348.
Hopkins, J. (1992). Psychoanalysis interpretation and science. In J. Hopkins & A. Saville (Eds.), *Psychoanalysis mind and art*. Oxford: Blackwell.
Kandel, E. R. (1999). Biology and the future of psychoanalysis: A new intellectual framework for psychiatry revisited. *American Journal of Psychiatry, 156*(4), 505–524.
Kellermann, N. P. F. (2011). Geerbtes Trauma: Die Konzeptualisierung der trans-generationalen Weitergabe von Traumata. In *Tel Aviver Jahrbuch für Deutsche Geschichte* (Vol. 39). Göttingen: Wallstein Verlag.
Klein, M. (1948). *Contributions to psychoanalysis 1921–1945*. London: Hogarth Press.
Laub, D. (2005). Traumatic shutdown of narrative and symbolization: A death instinct derivative? *Contemporary Psychoanalysis, 41*(2), 307–326.
Lester, B. M., Tronick, E., Nestler, E., Abel, T., Kosofsky, B., Kuzawa, C. W., Marsit, C. J., Maze, I., Meaney, M. J., Monteggia, L. M., Reul, J. M. H. M., Skuse, D. H., Sweatt, J. D., & Wood, M. A. (2011). Behavioral epigenetics. *Annals of the New York Academy of Sciences (Issue: Annals Meeting Reports), 1226*(1), 14–33.
Maquet, P., Peters, J.-M., Aerts, J., Delfiore, G., Degueldre, C., Luxen, A., & Franck, G. (1996). Functional neuroanatomy of human rapid-eye-movement sleep and dreaming. *Nature, 383*, 163–166.
Matsumoto, K., Suzuki, W., & Tanaka, K. (2003). Neuronal correlates of goal-based motor selection in the prefrontal cortex. *Science, 301*, 229–232.
McGowan, P. O., & Szyf, M. (2010). The epigenetics of social adversity in early life: Implications for mental health outcomes. *Neurobiology of Disease, 39*, 66–72.
McGowan, P. O., Sasaki, A., D'Alessio, A. C., Dimov, S., Labonté, B., Szyf, M., Turecki, G., & Meaney, M. J. (2009). Epigenetic regulation of the glucocorticoid receptor in human brain associates with childhood abuse. *Nature Neuroscience, 12*(3), 342–348.
Mishkin, M., & Appenzeller, T. (1987). The anatomy of memory. *Scientific American, 256*(6), 80–89.
Nofzinger, E. A., Mintun, M. A., Wiseman, M. B., Kupfer, D. J., & Moore, R. Y. (1997). Forebrain activation in REM sleep: An FDG PET study. *Brain Research, 770*, 192–201.
Northoff, G., & Hayes, D. J. (2011). Is our self nothing but reward? *Biological Psychiatry, 69*, 1019–1025.
Northoff, G., Wiebking, C., Feinberg, T., & Panksepp, J. (2011). The 'resting-state hypothesis' of major depressive disorder—A translational subcortical–cortical framework for a system disorder. *Neuroscience and Biobehavioral Reviews, 35*(9), 1929–1945. doi:10.1016/j.neubiorev.2010.12.007.
Palombo, S. R. (1973). The associative memory tree. *Psychoanalysis and Contemporary Science, 2*, 205–219.
Palombo, S. R. (1978). *Dreaming and memory: A new information processing model*. New York: Basic Books.

Panksepp, J. (1998). *Affective neuroscience: The foundations of human and animal emotions.* New York: Oxford University Press.
Piaget, J. (1928). *Judgment and reasoning in the child.* London: Routledge. (Original work published 1924)
Posner, M. I., & DiGirolamo, G. J. (1998). Executive attention: Conflict, target detection, and cognitive control. In R. Parasuraman (Ed.), *The attentive brain* (pp. 401–423). Cambridge, MA: MIT Press.
Reiser, M. F. (1984). *Mind, brain, body: Toward a convergence of psychoanalysis and neurobiology.* New York: Basic Books.
Reiser, M. F. (1990). *Memory in mind and brain: What dream imagery reveals.* New York: Basic Books.
Risch, N., Herrell, R., Lehner, T., Liang, K.-Y., Eaves, L., Hoh, J., Griem, A., Kovacs, M., Ott, J., & Merikangas, K. R. (2009). Interaction between the serotonin transporter gene (5-HTTLPR), stressful life events, and risk of depression: A meta-analysis. *The Journal of the American Medical Association, 301*(23), 2462–2471.
Roelofs, A., van Turennout, M., & Coles, M. G. H. (2006). Anterior cingulate cortex activity can be independent of response conflict in Stroop-like tasks. *Proceedings of the National Academy of Sciences, 103*, 13884–13889.
Solms, M. (2000). Dreaming and REM sleep are controlled by different brain mechanisms. *Behavioral Brain Sciences, 23*(6), 843–850.
Stuphorn, V., & Schall, J. D. (2006). Executive control of countermanding saccades by the supplementary eye field. *Nature Neuroscience, 9,* 925–931.
Winson, J. (1985). *Brain and psyche: The biology of the unconscious.* New York: Anchor.
Wollheim, R. (1999). *On the emotions.* New Haven: Yale University Press.

Part IV
ReVisions of the Drive in Freud and Neuroscience

Chapter 7
Beyond the Death Drive: Freud's Engagement with Cell Biology and the Reconceptualization of His Drive Theory

Sigrid Weigel

Abstract The chapter examines Freud's conception of drive defined by him as a limit concept (*Grenzbegriff*) between biology and the psyche. It further introduces this concept as an exemplary topic with which to study the relationship between natural science and cultural topics in Freud's theory. In analyzing Freud's reconceptualization of 'drive' in the last part of *Beyond the Pleasure Principle* (1920), the chapter concentrates on his references to biology and the organic, especially to the debate on the (im-)mortality of the organism in late nineteenth-century cell theory (Weismann et al.). Weigel shows how in Freud's drive theory the idea of the relation between 'mortal' and 'immortal' cells (cells of the organism and germ-cells) in biology get transferred into opposing forces, namely the 'tendency of the organic towards the an-organic' and the counter activity of Eros, i.e. the tension between death drive and life drive. The chapter concludes in summarizing the complex concept of 'life' developed in Freud's psychoanalysis.

Keywords Drive • Life • Death • Biology • Cell theory

7.1 The Drive on the Threshold Between Biology and the Psyche

The shaping of the concept of drive (*Trieb*) forms an intricate and complicated path within Sigmund Freud's writings. He introduced the word as one part of the compound term sexual drive (*Sexualtrieb*) in his *Three Essays on the Theory of Sexuality/Drei Abhandlungen zur Sexualtheorie* (1905) without offering a proper definition; instead he uses it in reference to a kind of bodily pressure (*Drang*) with various effects and impacts reaching far beyond the corporeal, for instance, when

---

S. Weigel (✉)
Zentrum für Literatur- und Kulturforschung Berlin,
Schützenstrasse 18, D-10117 Berlin, Germany
e-mail: direktion@zfl-berlin.org

the aim and the object of the sexual drive are at stake and when discussing the fantasies connected to different kinds of perversion. Within this context Freud's term drive possesses a strong sexual indexicality. As with the allegory that Plato uses in the *Symposium* to explain the origin of the sexes, in which male and female parts are described as two separated halves of a former male-female unity, the term drive itself seemed to be a divided term in search of sexual unification.

Ten years later, in *Instincts and their Vicissitudes/Triebe und Triebschicksale* (1915), Freud dedicated himself to explaining his idea of drive more explicitly and made an effort to finally reflect on the biological foundations of the drive: its force originates from what Freud calls the biological nature of its preconditions, that is, the nervous system's method of coping with stimuli (*Reizbewältigung*). With this process in place, he ascribes progress as fundamental to the drives within phylogenesis: they function as the motor behind the development of our highly evolved nervous system (Stud. III, 84). Here the term drive is freed from being a mere component of the sexual drive, as it takes on the form of an autonomous concept with a quite sophisticated structure. Now the drive appears as a liminal concept (*Grenzbegriff*) between the somatic and the soul, more precisely as "the representative of the stimuli arriving from the inner of the body and getting to the soul" (Stud. III, 85).[1] In other words, the drive is a representative of the somatic in the psychic apparatus; its preconditions are of a biological nature. Another five years later when Freud turned to the intense study of the biological preconditions of this liminal concept and situated it *Beyond the Pleasure Principle*, ending up with a much broader concept. Through a discussion of cell theory based on the biological knowledge of his time, he succeeded in developing the dialectical concept of the death drive and the life drive. He reconceptualized the sexual drive as the "true life drive" or Eros. As an energetic counterforce of the psychic apparatus, Eros works against the biological tendency of the organic to converge to the anorganic. In this dialectical conception Freud developed a synthesis of the aspects that he had analyzed in his earlier reflections on drives. The biological nature of the drives and their preconditions prompts the psychic apparatus to develop enormous energetic counterforces, forces of life that surpass mere biology; the character of this life is based in Eros, but it is more than sexual activity. It also provides the energetic basis for cultural achievements through the sublimation of the sexual drive.

Thus the term drive developed from part of a compound term via a liminal term to a dialectical constellation.[2] This complicated trace of the drive within Freudian theory makes the concept a fascinating example for his thinking on the threshold between the so-called two cultures as it continuously transgresses the lines that

---

[1] Own translation. Here and in the following quotations from Stud. III, I prefer my own translation of the German original instead of referring to the Standard Editions (S.E.), especially when the drive is at stake. Thus I want to avoid not only the mistranslation of *Trieb* as instinct, but also the displacement of connotations when the Freudian language gets translated into contemporary scientific terminology, as for example mental for psychic (*psychisch*) or soul (*Seele*) and organic for the inner of the body (*das Körperinnere*).

[2] As for the similarity to the dialectical structure of Walter Benjamin's theory see Weigel (2010a).

separate the two epistemologies. *Beyond the Pleasure Principle/Jenseits des Lustprinzips* (1920), a text that Sigmund Freud himself characterized as a metapsychological description, develops a concept of life that transcends the abiding antagonisms between the terminology of the natural sciences and the humanities.

In interpreting Freud's text as a component of knowledge of life beyond the 'two cultures' that remains controversial to this day, I will refer to some detours—detours in 'life' and detours in method. One might add as a supplement to the well-known Benjaminian adage "method is detour" (*Methode ist Umweg*, Benjamin 1972, 208) an idea from Freud's *Beyond the Pleasure Principle*: life too is a detour, a detour to death. This remarkable concept of life is found in the last part of the text, where Freud refers to phylogenetic theorems in order to explain that the "detours before reaching its aim of death" present us with "the picture of the phenomena of life" nowadays (S.E. XVIII, 39), before he ultimately converts this figure of the detour into the dynamic between the death and life drives and their different trajectories by way of the "hesitating rhythm" figure (*Zauderrhythmus*, Stud. III, 250).

In sections V and VI of *Beyond the Pleasure Principle* Freud develops his concept of life in his presentation of the debate on death and reproduction in the fields of evolutionary biology and medicine at the turn of the twentieth century, which leads him to ascribe an organic foundation to the philosophy of the will. The necessity "for borrowing from the science of biology" (S.E. XVIII, 60) prompts an expansion of the drive theory through which Freud anchors the work of the psychic apparatus as a kind of pressure (*Drang*) in the organic. In this way, the conception of drive is recast in the model of a dynamic counter constellation of the death and life drives. Against the background of its genesis in the history of science, this revised conception emerges as a construction into and in which older concepts, such as *Bildungstrieb* and sexual instinct, have been tacitly integrated and transformed in light of contemporary knowledge on the organism. My thesis is that the definition of "life-preserving Eros" as a force that opposes and works against the tendency of the organic to converge with the inorganic is the result of a psychoanalytical point of view, in which an interest in cultural questions leads to a detour that involves critical analysis of the natural sciences.

7.2 The Detour of Psychoanalysis

In 1935 Freud wrote a retrospective summary of his work in the *Postscript* to his autobiographical study (*Nachschrift zur Selbstdarstellung*[3]) that appeared in the *Almanach der Psychoanalyse* in 1936, 11 years after the *Autobiographical Study/Selbstdarstellung* (S.E. XX, 7–74) that described the development of his career in *Die Medizin der Gegenwart in Selbstdarstellungen*. From the vantage point of 1935, he views the theory of drives as one of two steps that concluded his psychoanalytical project. After putting forward the "hypothesis of the existence of

---

[3] In the English translation, this text is simply called *Postscript* (S.E. XX, 71–74).

two classes of drive (Eros and the death drive)" and proposing "a division of the mental personality into an ego, a super-ego, and an id (1923),"[4] Freud claims that he "made no further decisive contributions to psycho-analysis" (S.E. XX, 72).[5] But it would be wrong to infer from this statement that Freud did not appreciate his later works, which, after all, include such important texts as *Civilization and its Discontents/Das Unbehagen in der Kultur* (1930). Rather, he attributes this later text to a new endeavor: his attempt to use analytical insights in his investigation of "the origins of religion and morality" (S.E. XX, 72). He sees *Totem and Taboo* (1912) as a precursor to this, as it were, post-analytical complex of studies on the theory of culture. However, he makes no mention of the project on which he had just begun to work: *Moses and Monotheism/Der Mann Moses und die monotheistische Religion* (1939). The three treatises contained in that book would appear just three weeks after the outbreak of the Second World War in 1939, the same year that Freud died in exile in London.[6]

Written four years prior to his death, the *Postscript* is already characterized by a desire to take stock of the past. Freud talks about finishing his autobiographical notes and reflects on the progress made in the adoption of psychoanalysis in universities and medical praxis around the globe. In those parts of this text where he describes the reorientation of his interests towards cultural theory, Freud sketches a disconcerting image of his own scientific development: "My interest, after making a lifelong *détour* through the natural sciences, medicine and psychotherapy, returned to the cultural problems which had fascinated me long before, when I was a youth scarcely old enough for thinking" (S.E. XX, 72). Even if the longest phase of his scientific work—i.e., all the works culminating in the theory of drives and the topographical model of 'ego, super-ego and id' that mark the completion of psychoanalysis—is represented here as a long detour via the natural sciences and medicine, the transformation and reorientation of his work are nevertheless described as a "phase of regressive development" because the old Freud is returning to the themes of his youth.

A short passage from *An Autobiographical Study* (1925) reveals that the talk of a regression stems from an autobiographical perspective. Here we read that Freud's interest in questions of culture can apparently be traced back to a pre-scientific curiosity, which, though predating his studies, did prompt his decision in favor of medicine:

> Neither at that time, nor indeed in my later life, did I feel any particular predilection for the career of a doctor. I was moved, rather, by a sort of curiosity, which was, however, directed *more towards human concerns than towards natural objects*; nor had I grasped the importance of observation as one of the best means of gratifying it. (S.E. XX, 8, my italics)

---

[4] Freud is alluding to *The Ego and the Id* (1923).

[5] In the Standard Edition of the Complete Psychological Works of Sigmund Freud "Trieb" is translated as "instinct". In keeping with the current practice of referring to "drive" instead of "instinct", the following quotations from the Standard Edition have been modified accordingly.

[6] On September 23, 1939.

Freud claims, however, that Darwin's teachings[7] exerted a strong draw on him due to their promise of advancing the understanding of the world. Yet a lecture on *Nature/Die Natur*, an essay from the Goethe era, was the crucial factor in his decision to study medicine. Told *en passant*, this recollection reads like the foundational scene in psychoanalysis: the birth of medical curiosity from the lure of the aisthetic contemplation of the natural world, curiosity about human relations, and the promise of a greater understanding of the world. What Freud calls a regressive development in 1935 is therefore the return to a field that only now—after the completion of psychoanalysis—becomes accessible to him as an object of scientific investigation.

> I perceived ever more clearly that the events of human history, the interactions between human nature, cultural development and the precipitates of primaeval experiences (the most prominent example of which is religion) are no more than a reflection of the dynamic conflicts between the ego, the id and the super-ego, which psycho-analysis studies in the individual – are the very same processes repeated upon a wider stage (S.E. XX, 72).

In this way, he interprets his transfer of psychoanalytical theorems from the individual to the history of culture as a response to anthropological conflicts, whose origins lie in the interrelationship of nature, culture, and the after-effects of prehistoric experiences. The process of evolution described by Darwin is thus placed in that "primordial time" of which man has no recollections and the experiences of which are deposited in his memory—similar to an individual's early childhood experiences. If Freud's theory of culture is based on a phylogenetic model, then psychoanalysis as a point of view is positioned at the vanishing point where the indelible traces of the evolutionary pre-history of the species are expressed in the interaction of nature and culture, an anthropological unconscious as it were.

Freud's talk about the "phase of regressive development" in his work is illuminating, for it shows that the result of this regression is, as in practical analysis, a gain in knowledge, a kind of metapsychology of human nature. However, the fact that in the *Postscript* Freud categorizes all his work up to and beyond the mid-1920s as part work in the natural sciences and medicine is still a source of confusion. This contradicts the widespread view of Freud's course of development as a doctor and neurologist who, in a momentous methodological departure from his former goal of localizing psychical phenomena in the anatomical or physiological, invented a new science—psychoanalysis—in his forties and who then went on to concentrate on exploring the language of the unconscious, for example, the talking cure, the symptom in the *Studies on Hysteria* (1895b) written together with Breuer, and the modes of dream work in the *Interpretation of Dreams* (1900). Such an interpretation is supported not least by the aborted attempt in the *Project for a Scientific Psychology/Entwurf einer Psychologie* (1895a) "to furnish a psychology that shall be a natural science: that is,

---

[7] On the significance of Darwin for Freud, see Lucille Ritvo, *Darwin's Influence on Freud: A Tale of Two Sciences* (1990). Ritvo shows—not least with reference to several volumes in Freud's library with shelf marks from the years 1875 to 1883—that Freud studied Darwin's writings during his medical studies. She reads Freud's references to 'neo-Lamarckian' ideas in the light of his critical analysis of Darwin's work.

to represent psychical processes as quantitatively determinate states of specifiable material particles" (S.E. I, 295). This represented a methodologically advanced and epistemologically as yet unrealized attempt to link neurological processes and the significance of psychical mechanisms—in other words: to bridge the gap between the quantity and quality paradigms. It was to remain an unfinished project, which Freud is said to have forgotten later on. If, in the interpretive pattern of Freud's *Postscript*, everything that arose afterwards and in place of this project rooted in the natural sciences can still be ascribed to the natural sciences and medicine, his aforementioned *Autobiographical Study* challenges us to investigate more thoroughly the role the natural sciences played in the elaboration of psychoanalysis. If we take our cue from Freud that means that we ought to reconstruct the elaboration of psychoanalysis on its detour via the natural sciences and medicine. In what follows, I will explore this perspective in *Beyond the Pleasure Principle*, in particular in the last part on the theory of drives. In my view, the basic preconditions needed to reorient psychoanalysis towards questions of culture are developed there.

7.3 Beyond the Pleasure Principle: Discovering a "More Primordial" Drive

In the first section of *Beyond the Pleasure Principle* Freud formulates a precise task: the need to test an assumption that has been taken for granted in psychoanalytical theory that psychic events are regulated by the pleasure principle. This amounts to a questioning of the conceptual linking of sensations of pleasure and unpleasure with the quantity of excitations, "in such a manner that unpleasure corresponds to an *increase* in the quantity of excitation and pleasure to a *diminution*" (S.E. XVIII, 8). When Freud characterizes this assumption as uncritical (*unbedenklich*, Stud. III, 217) he alludes to the fact that too few thinkers have put their minds to it and thus invites a reevaluation of his energetic drive concept. His aim is, therefore, to test the assumption that the pleasure principle dominates and is derived from the constancy principle. In the process, Freud considers how the various sources of discharging unpleasure (*Unlustentbindung*) while at the same time pursuing a compatible metapsychological view, i.e., a description in which "we try to estimate this 'economic' factor in addition to the 'topographical' and 'dynamic' ones" (S.E. XVIII, 7).

However, Freud's text actually falls into two parts. And only the first part defers to the programmatic title, i.e., sections II, III, and IV, in which Freud outlines his widely discussed theory of trauma and investigates the effects of external shocks on the psychic apparatus. The symptoms of traumatic experiences such as shell shock after World War I and various manifestations of the repetition compulsion are points of departure for a revision of the theory of memory and the topography of the psychic apparatus. In the face of the "operation of tendencies *beyond* the pleasure principle" that Freud observes in the habit of "making what is in itself unpleasurable into a subject to be recollected and worked over in the mind" (S.E. XVIII, 17), he speculates that there is a disjunction between coming to consciousness (*Bewusstwerdung*) and leaving behind a memory trace (*Hinterlassung einer Gedächtnisspur*)—an assumption with radical consequences for any consciousness-based concept of the subject. In Freud's investigations of how the pleasure principle is suspended by a trauma acting from the outside, i.e., by external shocks that penetrate the protective shield, his description ultimately leads—in epistemological terms—to the prospect of an earlier, more primordial function of the psychic apparatus. This is considered to be the breakthrough to, as Freud puts it, the "prehistory" of the dream's wish-fulfilling tendency. With this prehistory, towards the end of section IV a vista opens onto a primordial history of human nature prior to wish fulfillment and the pleasure principle. A phylogenetic perspective thus comes into play. Yet Freud does not elaborate on it further here.

In section V, however, there is a change of topic—at least on the surface—as the text turns to examine the very different penetrations of the protective shield, namely, those shocks that derive from "internal excitation" rather than from external shocks: "The most abundant sources of this internal excitation are what are described as the organism's 'drives'" (S.E. XVIII, 34). The phenomenon of the repetition compulsion links both parts of *Beyond the Pleasure Principle* thematically. While in the first part Freud discusses the repetition compulsion provoked by traumatic experiences as a borderline case from an economic-neurological perspective on the psychic apparatus, the register changes in the second part. Here Freud investigates a more primordial repetition compulsion, which he ascribes to impulses that originate in drives. He calls this compulsion that precedes the pleasure principle the "organic compulsion to repeat" (S.E. XVIII, 37).

But the key to uncovering the more hidden link between the two parts is the question of the 'more primordial' (*das Ursprünglichere*). This leads beyond the title of the book: in methodological terms, to a sphere that completely transcends the parameters of pleasure-unpleasure back to what came before this parameter; and in terms of the subject matter, to a dynamic prior to the pleasure principle. Sections V and VI then address "ultimate things," those questions concerned with "the great problems of science and life" (S.E. XVIII, 59). Freud, however, divests the question of ultimate things (a pathos formula of countless philosophical and religious treatises) of its metaphysics and turns it back toward the organic. His discussion of inner excitations cedes almost imperceptibly to reflections on 'life,' more precisely on the "attribute of drives and perhaps of organic life in general" (S.E. XVIII, 36). If one attribute of drives is to allow us insights into organic life in general, then Freud hopes to gain these insights by asking how the drives are connected with the compulsion to repeat, a question to which he already has an answer: "*It seems, then, that a drive is an urge inherent in organic life to restore an earlier state* which the living matter has been obliged to abandon under the pressure of external disturbing forces; that is, it is a kind of organic elasticity, or, to put it another way, the expression of the inertia inherent in organic life" (S.E. XVIII, 36, transl. mod.).

With this shift in register from the neurological-economic *pleasure-unpleasure principle* to *organic life*, the meaning of 'repetition' also changes. While in the former, the repetition compulsion limits the validity of the pleasure principle and, therefore, also the tendency to reduce the tension of excitation, Freud conversely ascribes a persisting quality to the repetition in the organic—either as the expression of the conservative nature of the living, as the repetition of phylogenesis in the development of the germ cell, or in the form of a restoration of an earlier state, a kind of organic *restitutio in integrum*.[8] External actions are also given another role here: rather than compulsively setting repetition in motion, they provoke change because, according to Freud, we must attribute the achievements "of organic development [...] to external disturbing and diverting influences" (S.E. XVIII, 38). In this way, the second part of *Beyond the Pleasure Principle* goes beyond the neurological foundation of psychoanalysis and enters the arena of evolutionary biology. Yet Freud does not simply appropriate its theses, but outlines a concept of life—on a detour via the conservative character of the organic—in which Eros, the repression of drives, and sublimation pave human nature's way to culture, the counterforce to the death drive. The theory of drives is, therefore, not only an important step towards the completion of psychoanalysis, but also an essential prerequisite for a theory of culture—precisely because the theory of drives is based in the very bio-scientific debates that it will ultimately transgress.

To summarize the theoretical operation of his article: the epistemological step *beyond* the paradigm of pleasure and unpleasure leads, in terms of phylogenesis, to a drive concept prior to it. Against this background the sexual drive gets reformulated as a counterforce to the death drive that is inherent in the organic matter of the human body.

7.4 Beyond the Death Drive

In *Beyond the Pleasure Principle* Freud himself sees his drive theory model as the third step in his overall theory of drives after (1) broadening the meaning of the term sexuality (beyond the reproductive function) and (2) establishing narcissism as an ego-directed libido (S.E. XVIII, 59). The dynamic model that presents the opposition between the death drive and life drive stems from the critical analysis of a debate on the logic of death and reproduction in evolutionary theory, which had been raging since the 1880s among physicians and physiologists, including August Weismann, Alexander Götte, Max Hartmann, and others. Here Freud compares their theses with the suppositions of psychoanalytic libido theory. Yet, before describing and explaining this debate, Freud outlines his concept of the death and life drives in section V. The question of the 'more primordial' pursued here not only leads to the formulation of the drive concept as "*an urge inherent in organic life to restore an earlier state*" (S.E. XVIII, 36); it also results in an explicit naming of the more primordial: "'*inanimate things existed before living ones*'" (S.E. XVIII, 38). Consequently, the tendency of drives is characterized as conservative: "If we are to

---

[8] On Benjamin's appropriation and transformation of this idea into a "messianic nature", see Weigel (2010a).

take it as a truth that knows no exception that everything living dies for *internal reasons*—becomes inorganic once again—then we shall be compelled to say: *the aim of all life is death* and, coming back to: *inanimate things existed before living ones*" (S.E. XVIII, 38, transl. mod.).

The way in which Freud recounts the history of the—he emphasizes—unlikely sounding conservative nature of the drives is consistent with both evolutionary theory and the biblical formula: "Till thou return unto the ground; for out of it wast thou taken: for dust thou art, and unto dust shalt thou return" (Genesis 3:19). The point of departure is a kind of organic principle of constancy inherent in elementary living entities, which, Freud suggests, do not change under stable conditions, but "constantly repeat the same course of life" (S.E. XVIII, 38). In the short version of the phylogenetic history that Freud extrapolates from this, earth's historical development is the cause of changes that continue to shape the course of life for living beings and that are preserved in their drive to repeat. Hence the impression of change appears "whilst in fact they are merely seeking to reach an ancient goal by paths alike old and new," an initial state "from which the living entity has at one time or other departed and to which it is striving to return" (S.E. XVIII, 38).

In this narrative it is striking that an element of tension comes into play at the same time organic life arises—"at some time" traits of the living were awoken in inanimate substance. From his neurological-economic approach, Freud evidently ascribes this tension—via the concept of drive—to the idea of organic life: "The tension which then arose in what had hitherto been an inanimate substance endeavoured to cancel itself out" (S.E. XVIII, 38). As a result of "decisive external influences," the "surviving substance" was obliged to "diverge ever more widely from its original course of life and to make ever more complicated detours before reaching its aim of death" (S.E. XVIII, 38–39). Thus according to Freud, it is the "circuitous paths to death" that present us with a "picture of the phenomena of life" today (S.E. XVIII, 39). In any case, he suggests that one can reach no other conclusion if one assumes that the drives are wholly conservative in nature.

Freud's objection to this claim begins with a 'but'—"But let us pause for a moment and reflect. It cannot be so." (S.E. XVIII, 39)—and with this gesture he turns the spotlight onto the sexual drives and goes on to explain their role in an excursus on evolutionary biology. Without naming Weismann, Freud invokes the germ cells that differ from other elementary organisms in the way they detach themselves from the organism "with their full complement of inherited and freshly acquired aptitudes of drives," thereby counteracting the process of dying and attaining for us "what we can only regard as potential immortality, though that may mean no more than a lengthening of the path towards death" (S.E. XVIII, 40, transl. mod.). Here Freud alludes to the distinction drawn between germ and somatic cells in August Weismann's germ plasm theory. Contrary to Weismann's thesis of a *Continuity of the Germ Plasm as the Basis for a Theory of Inheritance/Continuität des Keimplasmas als Grundlage einer Theorie der Vererbung* (1885), in which germs are seen to be immutable vis-à-vis their environment, Freud suggests that they have the potential to transfer not just inherited but also acquired characteristics.

This conviction, which is often referred to as Freud's Lamarckism,[9] not only becomes important in the concept of phylogenetic memory that Freud later outlines in *Moses and Monotheism*;[10] it already plays a role in this text in the revision of the drive theory on an organic basis.

To the extent that Freud brings the germ cells into play as the counterpart to the mortality of the organic, they are consistent with the image of 'life as a detour to death.' The theoretically relevant intervention occurs once the group of sexual drives that precipitates the amalgamation of the germ cells is introduced, which, in turn, provides the cells with the capacity of transferal. By supplementing the germ cells with sexual drives, which enable the former to assume their function in the first place, Freud can affirm that the sexual drives "are the true life drives" (S.E. XVIII, 40). They retain life longer and thus counteract drives that lead to death. With the sexual drives, the concept of life shifts in *Beyond the Pleasure Principle* from its purely organic parameters (a detour to death) to an entire constellation of drives with opposing trajectories, whose dynamic Freud elucidates with a beautiful image:

> It is as though there exists a hesitating rhythm in the life of the organic; one group of drives rushes forward so as to reach the final aim of life as swiftly as possible; at a particular place of this path, the other group springs back to make the way from a certain point onwards once again and so prolong the way. (Stud. III, 250)

By imbuing 'life' with the sexual drive, Freud comes up with a 'solution' for the attempt to conceive of a transition from the organism to the psyche. The 'drive' enables the interconnection of the two registers and is, at the same time, the condition of possibility of *différance* as the "trace preceding the existing" (Derrida 1997, 47). In this way, Freud formulates his concept of life beyond the death drive, i.e., via the biological theorem of the "endeavour of all living substance [...] to return to the quiescence of the inorganic world" (S.E. XVIII, 62) and beyond.[11] At the same time, this theorem represents the matrix of his "speculation upon the life and death drives" (S.E. XVIII, 60), for Freud sees the urge to restore an earlier state of things as a general trait of the drives, not only the death drive, but also the life drive.

---

[9] This formulation is problematic because it implies an opposition between Lamarck and Darwin that, particularly with regard to the significance of external influences for species change, does not exist. On Darwin's Lamarckism, see the chapter on evolution and culture in my book: *Genea-Logik. Generation, Tradition und Evolution zwischen Kultur- und Naturwissenschaften* (2006). For an English translation of that chapter, see Weigel (2013). On the significance of epigenetic change for the theoretical conception of evolutionary biology, see Weigel (2010b).

[10] On Freud's critical reading of Lamarck and 'Lamarckism,' see Eliza Slavet (2008). Slavet does not just show that Freud was engaged in an intensive study of the ideas of Weismann and Lamarck from at least 1912 and planned to write a book on Lamarck together with Ferenczi. She also reveals just how much the scientific controversy on the inheritance of acquired characteristics was shaped by political constellations. One thinks immediately here of Weismann's proximity to the Gesellschaft für Rassenhygiene and the anti-Semitism inherent in the racial variant of a strictly genetic theory of inheritance and in the polemics that implied a connection between Judaism, psychoanalysis, and Lamarckism.

[11] On this figure, see Derrida (1980).

While the theory of drives masters the transition from organic to human nature in theoretical terms, this transition is represented through figurative language specific to psychoanalysis. Whenever Freud reflects on issues of language and representation, he considers figurative language as offering the only possible access to these problems:

> This is merely due to our being obliged to operate with the scientific terms, that is to say with the figurative language, peculiar to psychology (or, more precisely, to depth psychology). We could not otherwise describe the processes in question at all, and indeed we could not have become aware of them (S.E. XVIII, 60).

Apparently, in terms of perceptibility and describability, the gain outweighs the descriptive inadequacies, which, Freud remarks, would probably vanish if one were able to substitute them with physiological or chemical terminology. It seems not necessary to him to explicitly name the price that comes with using a nomenclature derived wholly from the natural sciences.

To summarize, on the trail of figurative language, the transformation of the concept of life from nature to culture is described in three stages, each identifiable as different periodic figures: *detour, hesitating rhythm, perfection by means of drive repression,* or more precisely, from the biological description of life as 'a detour to death,' via the 'hesitating rhythm in the life of the organism' effected by the sexual drive, to the perfection of culture as a product of drive repression. As for the latter, Freud repudiates the existence of a drive towards perfection that brought human beings "to their present high level of intellectual achievement and ethical sublimation" (S.E. XVIII, 42) with the argument that humans and animals have a comparable basis for development, i.e., the organic kinship of species as posited in evolutionary theory. Still, the sublimation thesis with which Freud contradicts the idea of a drive to perfection is nowhere to be found in evolutionary theory; besides, it would be difficult to apply this thesis to animals. In this way, *sublimation*, as a genuinely human capability, is given the status of a fundamental 'anthropological distinction.'[12] Freud interprets the restless urge to even greater perfection that he observes among a minority of people as the consequence of a drive repression,

> upon which is based all that is most precious in human civilization. The repressed drive never ceases to strive for complete satisfaction, which would consist in the repetition of a primary experience of satisfaction. No substitutive or reactive formations and no sublimations will suffice to remove the repressed drive's persisting tension; and it is the difference in amount between the pleasure of satisfaction which is *demanded* and that which is actually *achieved* that provides the driving factor which will permit of no halting at any position attained, but, in the poet's words '*presses ever forward unsubdued*' [Mephistopheles in Goethe's Faust, Part I, Scene 4] (S.E. XVIII, 42).

This alternative to the repudiated drive toward perfection, a dynamic driven forward by difference, is given the name *Eros* in the last paragraph of section V. Thus, unlike in the histories of humankind inspired by epigenesis a century before (in Herder, Blumenbach, Goethe and others), for Freud it is not perfectibility that is the driving force behind the history of culture, but life-preserving Eros, or the

---

[12] Here, I allude to the 'mosaic distinction.'

libido as the 'expression of the life drive.' That some elements from the history of epigenetics have nevertheless found their way into Freud's conception can be inferred from his discussion of biology and philosophy in section IV of *Beyond the Pleasure Principle*.

7.5 Life and Death

In this section Freud tests his claims regarding an inner law of dying against those explanations of "natural death" put forward by biology. He observes that biologists are greatly divided on this issue and refers to the evolutionary debate on the nexus of death and reproduction that went on for decades—a controversy that is surprisingly relevant for today's bio-demographic research and questions on the organic potential of extending the human lifespan. The debate was provoked at the end of the nineteenth century by the text *The Duration of Life/Ueber die Dauer des Lebens* (hereinafter referred to as DL) by August Weismann (1889a), the famous physician who, with his grounding of Darwin's evolutionary theory in cellular biology, is seen as having prepared the ground for neo-Darwinism. In this book, Weismann raises the question of the "minimum duration of life necessary for the maintenance of the species"(DL, 14). This leads him to the more general question of the reasons for death, which he judges to be "one of the most difficult problems in the whole range of physiology" (DL, 20).

In an attempt at an explanation in which he discusses the "necessity of reproduction" and the "utility of death," Weismann suggests that the "necessity of death" can only be understood in terms of usefulness (DL, 24). In this context, he deduces the emergence of "normal death, that is to say death which arises from internal causes" (DL, 27) from the transition from single-cell organisms that only multiply by cell division to higher developed organisms that multiply by sexual reproduction, and goes as far as attributing an eternal lifespan or immortality to primordial organisms. In this way, he interprets death as a form of evolutionary adaptation, which only first arises in the sexual reproduction of higher organisms. According to Weismann, while higher organisms lost the aptitude for eternal duration, a segmentation of their cells into two different kinds of cell groups had occurred: somatic cells (i.e., mortal and aging body cells) on the one hand and propagating germ or reproductive cells on the other. This division of labor means that all higher organisms still have an immortal core: the germ cell that guarantees heredity, survival and the persistence of the species. Yet, the imperative that a living creature survive the moment of reproduction at all is explained with reference to the care and rearing of its offspring, a development that also entails aging.

Above all, it was the thesis of potential immortality that prompted strong objections and controversial statements from contemporary scholars. The first written rebuttal was Alexander Göttes' (1883) *On the Origin of Death/Über den Ursprung des Todes*, in which he refutes Weismann's ideas on the continuity of life. In his desire for a clear distinction between the death of an entire organism and cell death, especially "postmortal cell death," Götte outlines an alternative "phylogeny of natural death" based on the counter-thesis that "reproduction is the exclusive and last reason for natural death" (Götte 1883, 32). In his riposte *Life and Death/Über Leben und Tod. Eine biologische Untersuchung,* Weismann reiterated his conviction and established it as a "fundamental biogenetic law" (1889b, 158). In accordance with this fundamental law, the connection between reproduction and death only applies to multicellular organisms, not to single-cell organisms. Here Weismann no longer just distinguishes between different cell types; rather, he distinguishes between the mortal and immortal halves of the individual: "The germ-cells are potentially immortal, in so far as they are able, under favourable conditions, to develop into a new individual, or, in other words, to surround themselves with a new body (*soma*)" (ibid., 122). More than a decade after Weismann's response, Max Hartmann (1906) attempted a synthesis in *Death and Reproduction/Tod und Fortpflanzung.* While he contradicts Weismann's thesis regarding the immortality of protozoa, he does accept his ideas on the continuity of the germ cell in multicellular organisms, even though he interprets death as an elementary occurrence that affects all organisms rather than a "state cultivated in the course of phylogenesis" (Hartmann 1906, 35). Although many individual issues were fought over unremittingly in this debate, their logic in evolutionary biology was increasingly reinforced, i.e., the thesis that the "question of the inner reason for natural death" coincided with the "question of the reasons for reproduction." For Hartmann, "death and reproduction are, in a sense, only the negative and positive aspects of the same problem, which is a problem of development" (ibid., 36).

Quite a few of the formulations and images of this debate are reflected in Freud's text, for example, Weismann's image of the germ cell that surrounds itself with a new soma—an image that is reminiscent of the Platonian concept of the soul that leaves the dying body in order to inhabit another. While such references to ancient knowledge are more likely to have slipped into the language of the cell theorist, Freud—in his use of figurative language for his psychoanalytical epistemology—uses them as the poet-philosopher's 'hint.' We see this again when he evokes Plato's myth of the genesis of the two sexes, the story of the two halves torn asunder that long to grow together, as a 'fantastic hypothesis.' He would not dare to refer to it here, "were it not that it fulfils precisely the one condition *whose fulfilment we desire*. For it traces the origin of a drive to a need to restore an earlier state of things" (S.E. XVIII, 57, my italics). His formulation alone suggests that curiosity also follows the same urge as the other life drives.

What has Freud gained from his reading of the biological debate? On the one hand, he adopts an organic matrix for his drive theory, in which the organic converges with the inorganic, drawing, for example, from Weismann reference to how "organic matter is continually passing, without residuum, into the inorganic" (DL, 34). On the other hand, he also integrates the idea of the division of living substance into soma and germ cell. It is this distinction that interested Freud the most in his reading of Weismann. He calls it an "unexpected analogy" to his own view, "which was arrived at along such a different path" (S.E. XVIII, 46). Yet, as distinct from the "living substance," Freud was more concerned with "the forces operating in it" and for that reason distinguished between two kinds of drives. When he describes his theory of drives as a "dynamic corollary to Weismann's morphological theory," (S.E. XVIII, 46) this confirmation of an affinity is, nevertheless, the point at which Freud begins to dissociate himself from the theoretical foundations of the entire debate. While he had previously expressed his opposition to Weismann's static heredity theory that rules out change in just a few words, he now emphatically distances himself from the biologist's logic of purposiveness. Freud claims that the controversy about the natural death of protozoa does not count at all for the questions that interest him; and he states that an understanding of sexuality in the sense of a "sober Darwinist way of thinking" (Stud. III, 265) contributes almost nothing to his own aims.

Contrary also to Jung's monistic libido theory (S.E. XVIII, 53), Freud presents an "exquisite dualistic view of the drives' life" (Stud. III, 258), in which the biological distinction between soma and germ cells is transformed into a dynamic concept of life with the help of the libido theory. In this concept, the death and life drives are translated into differentiable dynamics, as based on the characteristics of the two cell types that Weismann's germ plasm theory identified. It is clear from some of Freud's references that the tradition of epigenetic ideas played a role in this reworking. Among others, he looks to Ewald Hering (1878) to support his dualistic concept. According to Freud, Hering sees "two kinds of processes [...] operating in contrary directions" at work in living substance: a constructive/assimilatory process and a destructive/dissimilatory process (Stud. III, 49). And he goes on to write that he can make no secret of the fact that this journey led him unexpectedly into the harbor of Schopenhauer's philosophy: "For him death is the 'true result and to that extent the purpose of life', while the sexual drive is the embodiment of the will to live" (Stud. III, 50).

With this allusion, Freud refers to the theory of the will in Arthur Schopenhauer's (1909) *The World as Will and Representation/Die Welt als Wille und Vorstellung*—a theory that can indeed be read as a theory of drives—in particular to the fourth book "The Assertion and Denial of the Will to Live" in which Schopenhauer describes conception and death as manifestations of the will and the life of our body as "a constantly prevented dying, an ever-postponed death" (Schopenhauer 1909, 400). For Schopenhauer too, the will to life and death appear in the figure of an interlocked dynamic:

> [...] how the will to live in its assertion must regard its relation to death. We saw that death does not trouble it, because it exists as something included in life itself and belonging to it. Its opposite, generation, completely counterbalances it; and, in spite of the death of the individual, ensures and guarantees life to the will to live through all time. (Ibid., 424)

Schopenhauer arrived at this perception not least through his analysis of the epigenetic concept of *nisus formativus* or *vis formativis*, as conceptualized at the end of the eighteenth century and discovered by Wolff, Blumenbach, Herder, and others. Schopenhauer described this inherent force in the germ, the *Bildungstrieb* as the objectified will of living entities (ibid., 145). While Freud grounds Schopenhauer's concept of the will in biology, he conversely inscribes a dynamic into the biological knowledge of the organism. In psychoanalytical theory, this dynamic is called the pleasure principle because what Freud perceives as beyond the death drive is, in fact, the pleasure principle or the libidinous character of the drives. While the first part of *Beyond the Pleasure Principle* involves unearthing the death drive—more precisely, the organic basis of the drives and their tendency to converge with the inorganic, the second part introduces the life drive or Eros beyond the death drive. In this section Freud rediscovers the pleasure principle and presents a reconceptualization of the concept: "The pleasure principle seems to act in the service of the death drive" (Stud. III, 272).

7.6 Reintroducing the Question of Quality

The conception of the death and life drives and the dialectic of sublimation that produces culture are thus important milestones on the way to Freud's later writings; they are prerequisites for the formulation of his theory of culture. By inscribing tensions into the organic, Freud found a way to differentiate the 'inner excitations.' In working out the metapsychological aspects of his theory of drives and its reformulation at the limits of the organic and Eros, Freud also came up with a possibility of dealing with the gap between physiological and psychical phenomena, namely, by introducing a quality that lies prior to every ordained meaning, prior to an evaluative semantics or any kind of semantics for that matter. This can be read as a belated echo of the forgotten, unresolved question posed many years before in the *Project of a Scientific Psychology*.

From this perspective, Freud's work on the psychoanalysis of culture can indeed be understood as a detour via the natural sciences, not just by way of an explicit engagement with bio-scientific theorems in part two of *Beyond the Pleasure Principle*, but also by recovering unfinished traces from the *Project* written 25 years before, as with the question of "*where* qualities originate," which Freud raises in section 7 on "The Problem of Quality." In that section, Freud considers the role of every psychological theory "apart from what it achieves from the point of view of natural science," asking how contents can be arranged among quantitative processes, how quantities become qualities, etc. (S.E. I, 307). Unlike with neurons, which strive towards a *restitutio in integrum* like the organs of perception, Freud believed that facilitations (*Bahnungen*) had to arise from something other than quantities. Back then, he saw a solution in the temporal nature of the processes, in the periods of excitation that he viewed as the foundation of consciousness. He suggested that the qualities of sensations resulted from the diversity of the periods. At that time Freud tried to find a way around this insufficient explanation by introducing the sensations of pleasure-unpleasure—without actually coming up with a solution.

Beyond the Pleasure Principle begins precisely at the point where the *Project of a Scientific Psychology* left off.[13] Towards the end of this essay, Freud again tries to solve this problem by attempting to transfer "the libido theory which has been arrived at in psycho-analysis to the mutual relationship of cells" (S.E. XVIII, 50). By

---

[13] Still, the motif of the zone of indifference between pleasure and unpleasure in the *Entwurf*, which is taken up again in section I of *Beyond the Pleasure Principle* and attributed to Fechner, plays a less significant role here. The problem of integrating qualitative concepts into quantitative models is more important.

reintroducing the pleasure principle beyond the death drive, Freud aims to inscribe elements of *difference* and the work differentiation into a perspective rooted in the natural sciences. He does so by merging the constancy principle or the principle of an equalization of tension with the organic urge to restore an earlier state. We see this, for example, in the "change in the magnitude of the cathexis *within a given unit of time*" (S.E. XVIII, 63) of the pleasure-unpleasure series, which is what he calls the excitations that come from within at the end of the text. He comes to the conclusion already cited above: "The pleasure principle seems actually to serve the death drives" (S.E. XVIII, 63). However, the passage through biological knowledge does not leave the theory of drives unchanged: "The distinction between the two kinds of drive, which was originally regarded as in some sort of way *qualitative*, must now be characterized differently—namely as being *topographical*" (S.E. XVIII, 52).

With regard to philosophy, which traditionally claims to clarify the ultimate things, Freud thus revisited a perspective he had formulated much earlier in the *Psychopathology of Everyday Life* (1901): "One could venture to explain in this way the myths of paradise and the fall of man, of God, of good and evil, of immortality, and so on, and to transform *metaphysics* into *metapsychology*" (S.E. VI, 259). At the same time, Freud undertook a remarkable and momentous reversal of perspective vis-à-vis the biological knowledge of life and death. While Weismann and his colleagues looked for the cause of natural death in reproduction, Freud understands Eros and the dynamic of the living as a countermovement to mortality or to the striving of the organism towards death.

7.7 Postscript: "What Is Life?"

Twenty-three years later, Erwin Schrödinger (1944) would ask in his Dublin Lecture *What is Life?*—considered the primal scene of the search for the genetic code—just what it is in the organism that counteracts the tendency of matter towards heat death, towards entropy. The search for an answer to this question led to work on deciphering the genetic code. This work however resulted in a theory of evolution that is grounded in genetics and armed with molecular biology as well as in a sort of life sciences that has lost sight of the concept of 'life.' But it hasn't been until just recently that scholars in some neuro-scientific labs and many psychiatric clinics have turned their attention back to this concept and have been rediscovering the very questions Freud formulated a century ago.

References

Benjamin, W. (1972). Erkenntniskritische Vorrede. In R. Tiedemann & H. Schweppenhäuser (Eds.), *Gesammelte Schriften* (Vol. I, p. 208). Frankfurt am Main: Suhrkamp.
Derrida, J. (1980). *The post card: From socrates to Freud and beyond* (A. Bass, Trans.). Chicago: University of Chicago Press.

Derrida, J. (1997). *Of grammatology* (G. C. Spivak, Trans.). Baltimore: John Hopkins University. (Original work published 1967)
Freud, S. (1895a). *Project for a scientific psychology*. S.E. I.
Freud, S. (1895b). *Studies on hysteria*. S.E. II.
Freud, S. (1900). *Interpretation of dreams*. S.E. IV.
Freud, S. (1901). *The psychopathology of everyday life*. S.E. VI.
Freud, S. (1905). *Three essays on the theory of sexuality*. S.E. VII.
Freud, S. (1912). *Totem and taboo*. S.E. XIII.
Freud, S. (1915). *Triebe und Triebschicksale/Instincts and their Vicissitudes*. Stud. III./S.E. XIV.
Freud, S. (1920). *Beyond the pleasure principle*. S.E. XVIII.
Freud, S. (1923). *The ego and the id*. S.E. XIX.
Freud, S. (1925). *An autobiographical study*. S.E. XX.
Freud, S. (1930). *Civilization and its discontents*. S.E. XXI.
Freud, S. (1936). *Postscript*. S.E. XX.
Freud, S. (1939). *Moses and monotheism*. S.E. XXIII.
Götte, A. (1883). *Über den Ursprung des Todes*. Hamburg: L. Voss.
Hartmann, M. (1906). *Tod und Fortpflanzung*. Munich: Ernst Reinhardt.
Hering, E. (1878). *Zur Lehre vom Lichtsinn*. Vienna: Gerold.
Ritvo, L. (1990). *Darwin's influence on Freud: A tale of two sciences*. New Haven: Yale University Press.
Schopenhauer, A. (1909). *The world as will and idea* (R. B. Haldane & J. Kemp, Trans.). London: Kegan Paul, Trench, Trübner & Co. (Original work published 1818/1819)
Schrödinger, E. (1944). *What is life?* Cambridge: Cambridge University Press.
Slavet, E. (2008). Freud's 'Lamarckism' and the politics of racial science. *Journal for the History of Biology, 41*, 37–80.
Weigel, S. (2006). *Genea-Logik: Generation, Tradition und Evolution zwischen Kultur- und Naturwissenschaften*. Munich: Fink.
Weigel, S. (2010a). Treue, Liebe, Eros: Benjamins Lebenswissenschaft. *DVjSch, 4*, 580–596.
Weigel, S. (2010b). An der Schwelle von Kultur und Natur: Epigenetik und Evolutionstheorie. In V. Gerhardt & J. Nida-Rümelin (Eds.), *Evolution in Natur und Kultur* (pp. 103–123). Berlin: De Gruyter.
Weigel, S. (2013). The evolution of culture or the cultural history of the evolutionary concept: Epistemological problems at the interface between the two cultures. In B. Larson & S. Flach (Eds.), *Darwin and theories of aesthetics and cultural history* (pp. 83–107). Farnham: Ashton.
Weismann, A. (1885). *Continuität des Keimplasmas als Grundlage einer Theorie der Vererbung* [Continuity of the germ plasm as the basis for a theory of inheritance]. Jena: Fischer.
Weismann, A. (1889a). The duration of life. In A. E. Shipley (Trans.), *Essays upon heredity and kindred biological problems* (pp. 1–65). Oxford: Clarendon. (Original work published 1882)
Weismann, A. (1889b). Life and death. In A. E. Shipley (Trans.), *Essays upon heredity and kindred biological problems* (pp. 107–160). Oxford: Clarendon. (Original work published 1884)

Chapter 8
Drive and Love: Revisiting Freud's Drive Theory

Yoram Yovell

Abstract In light of recent neuroscientific findings relevant to psychoanalytic theories about drive and love, Yovell addresses two interrelated and partially overlapping questions: What can the neurosciences contribute to a psychoanalytic understanding of drive? And what can they contribute to a psychoanalytic understanding of how romantic love, sexuality and attachment relate to each other? The answers that Yovell provides unfold in his presentation of Freud's drive theory and John Bowlby's attachment theory, which he backs with neuroscientific research findings. The author also examines how the neuroscientist Jaak Panksepp's SEEKING emotional system relates to Freud's concept of libido, while the systems Panksepp named PANIC/GRIEF and CARE correlate with Bowlby's attachment system. Taking the example of adult romantic love, Yovell describes how Panksepp's subcortical emotional systems influence our conscious experience and our actions in an intricate interplay that remains to be thoroughly examined. In the end, the author conceptualizes drive theory and attachment theory as building on one another rather than as being mutually exclusive or otherwise contradicting each other.

Keywords Drive theory • Attachment theory • John Bowlby • Jaak Panksepp • SEEKING • PANIC/GRIEF • CARE • Love • Sexuality • Libido • BBURP

8.1 Drive and Love in Freud and More Recent Revisions: Bowlby and the Neurosciences

Throughout his life, Freud sought to investigate the motivations underlying human relationships, especially love relationships. His theories about the mental forces and developmental trajectories that lead to adult romantic love as described in his theory of sexuality and his essay on the general abjection of love (*Three Essays on the*

The author wishes to acknowledge the contribution of Mark Solms to the discussion of the psychoanalytic and neurobiological conceptualizations of drive.

Y. Yovell (✉)
Institute for the Study of Affective Neuroscience (ISAN), University of Haifa,
Haifa 31905, Israel
e-mail: yovell@research.haifa.ac.il

Theory of Sexuality/Drei Abhandlungen zur Sexualtheorie, 1905; *On the Universal Tendency to Debasement in the Sphere of Love/Über die allgemeinste Erniedrigung des Liebeslebens*, 1912) are, like his theories on repression (*Repression/Die Verdrängung*, 1915a), among the foundations of psychoanalytic thought. Throughout the last century his drive-based theories have served as a basis for psychoanalytic investigation and understanding of human development and human relationships in general (Kernberg 1995; Mitchell 1997; Yovell 2008).

Drive theory, which has become less popular in contemporary psychoanalysis, was formulated as a systematic attempt to describe the most basic motivational determinants of desire, emotion, thought and behavior from the perspective of human subjective experience (Freud's *Instincts and their Vicissitudes/Triebe und Triebschicksale*, 1915b; Mitchell and Black 1995; Schmidt-Hellerau 2001). Even contemporary psychoanalytic theorists who reject part or all of Freud's drive theory acknowledge its importance as a starting point for a psychoanalytic discussion of human motivation (Apfelbaum 2005).

The work of John Bowlby who was concerned mainly with the mother-child-relationship (1958, 1988) and other attachment theorists provide a different, and in many ways complementary, view about the origins of the human need to love and be loved (reviewed in Hazan and Shaver 1987). This has created a tension between attachment theories and drive-based theories of human motivation. Some contemporary attachment theorists have gone as far as rejecting the Freudian concept of drive altogether (Mikulincer and Shaver 2007). Others have argued that drive theory, perhaps in conjunction with attachment-based models, might still be useful for the understanding of human motivations (Diamond and Blatt 2007). Interestingly, there is a growing tendency for a rapprochement between psychoanalytic theory and attachment theory as for example in Peter Fonagy's approach of mentalization (Fonagy 2001).

Recent advances in the cognitive and affective neurosciences have enabled psychologists and neurobiologists to investigate different aspects of romantic love, including its motivational basis (Aron et al. 2005; Fehr 2001; Meyers and Berscheid 1997). There have also been numerous studies that explored the underlying neural mechanisms of sexual arousal, sexual activity and pair formation (Beauregard et al. 2001; Insel et al. 1998; Karama et al. 2002; Tiihonen et al. 1994; Van Goozen et al. 1997; reviewed in Fisher 2004). These developments have rekindled the scientific and psychoanalytic interest in the concept of drive in general and libido in particular (Pfaff 1999; reviewed in Yovell 2008).

8.2 Neuroscientific Contributions to the Understanding of Drive and Love

In light of the recent neuroscientific findings relevant to psychoanalytic theories about drive and love, I would like to examine two interrelated and partially overlapping questions: (1) What can the neurosciences contribute to our psychoanalytic understanding of drive? (2) What can the neurosciences contribute to our psychoanalytic understanding of how romantic love, sexuality and attachment relate to each other?

8.2.1 The Concept of Drive

Above and beyond its relevance for describing and defining the origins of love, the concept of drive is fundamental for our understanding of the functioning of the mental apparatus as a whole (Schmidt-Hellerau 2001). Drive is the most basic motivational concept, and therefore the most elementary concept in explaining desire, emotion, thought and behavior from the perspective of human subjective experience.

It is probably no coincidence that Freud struggled with the concept of drive as he attempted to construct a coherent psychological theory about the structure and function of the human mental apparatus, as may be seen in *Instincts and their Vicissitudes* (1915b). Indeed, a major stumbling block for the psychoanalytic understanding of drive as a source of motivation is the fact that it is as distant from introspective awareness as is possible. As defined by Freud, drive is the most "biological" or embodied part of the mind, and therefore the farthest removed from its conscious surface.

8.2.2 Why the Neurosciences Might Be Helpful in Refining Our Psychoanalytic Conceptualization of Drive

As Mark Solms and Edward Nersessian in their questioning of *Freud's Theory of Affect* from a neuro-scientific perspective (1999) pointed out, it is very difficult to investigate, classify and define such a basic phenomenon psychoanalytically when the observational data consists of statements such as "I feel like this or that." It is possible that Freud's difficulties with the concept and definition of the drives, as well as the lack of popularity of drive theory in the work of many contemporary psychoanalytic theoreticians (Apfelbaum 1966, 2005; Mitchell and Black 1995) may stem from this formidable problem of observation.

While the psychoanalytic observational perspective is very distant from basic bodily processes, the opposite is true for the neurosciences, which are therefore well-suited for mapping out the relationship between the organ of the mind and the rest of the body. Thus, the neurosciences may become helpful in defining the drives psychoanalytically.

8.2.3 Freud's Drive Theory: Between Mind and Brain

Freud was explicit when he stated that drive is neither a mental nor a biological concept. In his view, drive is a frontier element between body and mind. Rather than defining it as part of either one, the Freudian definition of drive has to do with the relationship between the two: how the body becomes mentalized and how it becomes represented in the mind. As a corollary, since drive is not mental, it can never be conscious and can never be experienced (Freud 1915b).

Therefore, when a person says "I feel driven," this is not a direct experience of a drive because drive is prior to all mental things. The analogy of gravity that Pfaff et al. (2007) have suggested may be useful here in clarifying the relationship between drive and conscious experience: Like gravity, drive is an inference. As an inferred force, it is never something you feel or see, but rather something that explains and predicts the sensations that you feel and the phenomena that you see (Pfaff et al. 2007).

8.2.4 The Driven Brain: Where Should We Look?

The obvious place to look for the site where the body's demands become mentalized would be the hypothalamus, and, more specifically, the hypothalamic "need detectors" (Panksepp 1981; Solms and Turnbull 2002). These homeostatic control centers register the state of various bodily physiological processes and are the sites in which bodily needs are represented in the brain. Their activation evokes strong subjective sensations of hunger, thirst, sexual arousal, etc. The influences of these control centers are not transmitted exclusively (or even primarily) by neural pathways but rather by more diffuse mechanisms (hormones, etc.). The psychoanalytic concept of drive has an energetic aspect and implies a mechanism for activation or arousal of the mental apparatus.

Interestingly, such a basic mechanism for generalized brain arousal exists in all vertebrates, including humans. Donald Pfaff (1999), who in his laboratory of neurobiology and behavior studied and characterized this system, nicknamed it BBURP (Bilateral, Bipolar, Universal Response Potentiating System). From a psychoanalytic perspective, it supplies the psychic energy that is required for all motivated thought, emotion and behavior. BBURP originates in the phylogenetically ancient reticular formation of the medial and ventral brainstem and projects both upwards and downwards (hence "bipolar"). BBURP system neurons send out ascending axons that potentiate sensorimotor and emotional aspects of brain responsiveness as well as descending axons that potentiate autonomic aspects of brain responsiveness (Pfaff et al. 2007).

8.2.5 Moving Upwards: Libido and SEEKING

BBURP supplies the mental energy that drives all emotional systems and processes, and indeed consciousness itself. One (and only one) of these systems is the ascending dopaminergic mesolimbic and mesocortical pathway (Panksepp 1998; Pfaff et al. 2007). Like other emotional command systems, it is "limbic." The ascending mesolimbic and mesocortical pathway is the emotional command system that Jaak Panksepp called SEEKING (Panksepp 1998, 2003). Its main neurotransmitter is dopamine (Panksepp 2006). SEEKING is strikingly similar to Freud's concept of libido. It underlies our desire to go out into the world and seek in it things that will supply our needs and give us pleasure. It underlies our sexual urges; it is important for dreams; it is overactive in psychosis and mania; and it can be activated by cocaine (Panksepp 2005, 2006; Solms and Turnbull 2002).

Freud had different views at different times about how to classify the drives (Schmidt-Hellerau 2001). But all the different ways in which he divided and categorized the drives that are relevant for our discussion were always subsumed under the single heading of libido, Eros or the life drive (Freud 1915b; *Beyond the Pleasure Principle/Jenseits des Lustprinzips*, 1920; *An Outline of Psycho-Analysis/Abriß der Psychoanalyse*, 1940). All the demands that are, as Freud writes in *Instincts and their Vicissitudes,* "made upon the mind for work in consequence of its connection with the body" (S.E. XIV, 122), all of our different appetites, when taken together, comprise this broad libidinal drive. To Freud, all appetites were, in the final analysis, sexual (Freud 1940).

8.3 Can We Have Motivations That Are Independent of Libido?

As will be detailed below, the Freudian idea that libido is the sole motivator of thought and behavior is probably incorrect. There is considerable biological evidence to suggest that, even putting "Thanatos" aside, not all our motivations may be reduced or related to libidinal wishes (Yovell 2008).

One important reason we can say this is because at the same neural level as the SEEKING system, there exist other emotional command systems that can function independently of SEEKING (Panksepp 1998; Panksepp and Biven 2012). This is where Bowlby enters the picture (Mikulincer and Shaver 2003, 2007).

8.3.1 Bowlby and Attachment

John Bowlby, who was trained as a psychoanalyst, hypothesized the existence of an inborn instinctual attachment system that "served the function of binding the child to its mother and which contributed to the reciprocal dynamic of binding mother to child" (Bowlby 1958, 351). He based his reasoning on clinical observations, psychoanalytic object relations theory and data derived from ethological research on primates. Importantly, he emphasized that the attachment system was not dependent on drive reduction and that it was not mediated through infantile sexuality (Bowlby 1988; Mikulincer and Shaver 2003, 2007).

8.3.2 Neural Correlates of Bowlby's Attachment System: Panksepp's PANIC/GRIEF and CARE Systems

Building on his own work as well as the work of other affective neuroscientists, Jaak Panksepp described an emotional command system that he called the PANIC/GRIEF system (Panksepp and Biven 2012). Like the SEEKING system, it is "limbic" but phylogenetically younger than SEEKING, having evolved in mammals and birds, i.e., animals that nurture and take care of their young. Its main anatomical locus is the anterior cingulate cortex (ACC), with ascending fibers into the cerebral cortex and descending fibers that feed into the periaqueductal gray (PAG).

From an evolutionary and behavioral standpoint, the initial role of the PANIC/GRIEF system was probably to bind the infant to its mother by triggering separation distress (in both of them) as they moved apart from each other. The important neurotransmitters and neuromodulators involved in this system are serotonin, oxytocin and endorphins (Panksepp and Biven 2012). This system is active in normal sadness, and is dysregulated and overactive in human depression (Panksepp and Watt 2011).

The CARE system, which has also been called the Tend-and-Befriend system (Taylor et al. 2000), is also characteristic of mammals and birds. It is activated in females before they give birth; it is responsible for maternal caretaking behavior; it exists in males as well, and is involved in paternal care; and, like the PANIC/GRIEF system, it is relatively independent of the SEEKING system (reviewed in Panksepp and Biven 2012). The CARE system is mediated by endorphins and oxytocin, as well as by other neuropeptides (Insel 2003). It appears that this system mediates the motivation for the intimate bonding between mother and infant that is not triggered by separation distress. It is likely that the CARE system underlies human friendship and bonding with peers, not only with offspring (Taylor et al. 2000). As such, it is an important foundation of the complex and heterogeneous mental state we call love.

8.3.3 Why Love Hurts: The PANIC/GRIEF System and Separation Distress

150 million years of natural selection have strongly favored those mothers and infants who tended to stick together (reviewed in Yovell 2008). Thus, by a relentless process of natural selection, separation distress has evolved to become an exceptionally powerful alarm system that signals infants and mothers that they have moved away from each other and motivates them to return to each other's proximity as soon as possible.

The activation of this system is subjectively painful. The neural pathways that mediate physical pain and those that mediate the emotional pain of social loss, rejection and exclusion partially overlap here (Eisenberger and Lieberman 2004). Moreover, it has recently been found that viewing pictures of attachment figures when being separated from them decreases the sensation of physical pain, decreases activity in pain-related neural regions and increases activity in safety-related neural regions (Eisenberger et al. 2011).

Whereas the dominant neurotransmitter in the SEEKING system is dopamine, the dominant neurotransmitters in the PANIC/GRIEF system are endorphins, as well as other neuropeptides and serotonin. The mechanism by which social abandonment, loss or rejection triggers mental pain is probably related to cessation of endorphin release in limbic brain regions. The fact that endorphins relieve both physical and social pain explains why narcotics, which are agonists of endorphin receptors, are so addictive (Insel 2003; Panksepp 1998), and why we are in such pain when we are abandoned by our loved ones (Panksepp and Biven 2012).

This inevitable, hard-wired connection between social loss and mental pain was well known to Freud, who in *Civilization and its discontents/Das Unbehagen in der Kultur* (1930) wrote:

> What is more natural than that we should persist in looking for happiness along the path on which we first encountered it? The weak side of this technique of living is easy to see; otherwise no human being would have thought of abandoning this path to happiness for any other. It is that we are never so defenseless against suffering as when we love, never so helplessly unhappy as when we have lost our loved object or its love. (S.E. XXI, 82)

8.3.4 Frantic Efforts and Insecure Attachments

An extreme example of how fears of abandonment have the power to elicit emotional turmoil and clinging/pleading behaviors is the syndrome of Borderline Personality Disorder (Gunderson 1996). This association is evident in Criterion 1 (i.e., most prevalent) of the DSM-IV-TR definition of Borderline Personality Disorder: "Frantic efforts to avoid real or imagined abandonment" (American Psychiatric Association 2000, 710). Borderline Personality Disorder has been associated with abnormal neurotransmission of endorphins, oxytocin and vasopressin (Stanley and Siever 2010). These neurotransmitters and neuromodulators are involved in the action of the PANIC/GRIEF and CARE emotional command systems (Panksepp and Biven 2012).

From the perspective of attachment theory, Borderline Personality Disorder is an extreme case of a dependent/fearful/preoccupied attachment style in adulthood (Agrawal et al. 2004). It has been found that attachment style in early childhood is highly predictive of attachment style in adulthood (Mikulincer 1998; Mikulincer and Shaver 2003). Thus, it appears reasonable that a childhood history of insecure love attachments, which has been found to have biological correlates, is a vulnerability marker for the later development of Borderline Personality Disorder.

8.3.5 Baby Love and Adult Love—How Similar Are They?

The evidence for the persistence of attachment style and attachment vulnerabilities from childhood to adulthood is in agreement with the similarities between the neuro-biological correlates of parental and romantic love. In an fMRI study, Andreas Bartels and Semir Zeki (2004) have shown that there are strong similarities between the patterns of regional activation in the brains of parents looking at a photograph of their child and lovers looking at a photograph of their loved one. Moreover, there are striking similarities between the patterns of brain deactivation (i.e., brain areas that become less active) in parents looking at a photograph of their child and lovers looking at a photograph of their loved one.

Interestingly, the areas that are deactivated in both cases are the tertiary, heteromodal cortical areas—developmentally the most recent, sophisticated areas of the brain, which are responsible for abstract thought, planning and reasoning (Bartels and Zeki 2004). The finding that the brain regions that mediate our highest cognitive functions are relatively inactive when people are thinking of their loved ones, either child or lover, is in agreement with the common wisdom that "love is blind," which is to say it is irrational.

In addition, the many anatomical, neurochemical and behavioral similarities that have been found between adult romantic love and infant-mother love support Freud's and Bowlby's hypothesis that adult romantic love is built on the foundations of the child's earliest attachments to those he loves.

8.4 In Conclusion

We have seen that adult romantic love is based on the action of (at least) two distinct and powerful limbic emotional and motivational systems: (1) The SEEKING system that mediates sexuality, attraction, excitement, novelty, etc.; the action of this system resembles the Freudian concept of *libido*; (2) The PANIC/GRIEF and CARE systems that mediate attachment, bonding, nurturance, separation distress, etc., and are relatively independent of SEEKING/libido.

These emotional command systems are ancient and exist in all mammals and birds. Importantly, they exist in all humans. The vicissitudes of the two systems' interplay have kept lovers, poets and psychoanalysts very busy for as long as we can remember and are likely to continue to do so in the future.

References

Agrawal, H. R., Gunderson, J., Holmes, B. M., & Lyons-Ruth, K. (2004). Attachment styles with borderline patients: A review. *Harvard Review of Psychiatry, 12*, 94–104.
American Psychiatric Association. (2000). *Diagnostic and statistical manual of mental disorders, fourth edition, text revision*. Washington, DC: American Psychiatric Association.

Apfelbaum, B. (1966). On ego psychology: A critique of the structural approach to psycho-analytic theory. *International Journal of Psycho-Analysis, 47*, 451–475.
Apfelbaum, B. (2005). The persistence of layering logic: Drive theory in another guise. *Contemporary Psychoanalysis, 41*, 159–181.
Aron, A., Fisher, H., Mashek, D. J., Strong, G., Li, H., & Brown, L. L. (2005). Reward, motivation, and emotion systems associated with early-stage intense romantic love. *Journal of Neurophysiology, 94*, 327–337.
Bartels, A., & Zeki, S. (2004). The neural correlates of maternal and romantic love. *NeuroImage, 21*, 1155–1166.
Beauregard, M., Levesque, J., & Bourgouin, P. (2001). Neural correlates of conscious self-regulation of emotion. *Journal of Neuroscience, 21*(RC165), 1–6.
Bowlby, J. (1958). The nature of the child's tie to his mother. *International Journal of Psycho-Analysis, 39*, 350–373.
Bowlby, J. (1988). *A secure base: Clinical applications of attachment theory*. London: Routledge.
Diamond, D., & Blatt, S. J. (2007). Editors' introduction. In D. Diamond, S. J. Blatt, & J. D. Lichtenberg (Eds.), *Attachment and sexuality* (pp. 1–26). New York: Analytic Press.
Eisenberger, N. I., & Lieberman, M. D. (2004). Why rejection hurts: A common neural alarm system for physical and social pain. *Trends in Cognitive Sciences, 8*, 294–300.
Eisenberger, N. I., Master, S. L., Inagaki, T. K., Taylor, S. E., Shirinyan, S., Lieberman, M. D., & Naliboff, B. D. (2011). Attachment figures activate a safety signal-related neural region and reduce pain experience. *Proceedings of the National Academy of Sciences, 108*, 11721–11726.
Fehr, B. (2001). The status of theory and research on love and commitment. In G. Fletcher & M. Clark (Eds.), *Blackwell handbook in social psychology: Vol. 2. Interpersonal processes* (pp. 331–336). Oxford: Blackwell.
Fisher, H. (2004). *Why we love*. New York: Holt.
Fonagy, P. (2001). *Attachment theory and psychoanalysis*. New York: Other Press.
Freud, S. (1905). *Three essays on the theory of sexuality*. S.E. VII.
Freud, S. (1912). *On the universal tendency to debasement in the sphere of love (Contributions to the psychology of love II)*. S.E. XI.
Freud, S. (1915a). *Repression*. S.E. XIV.
Freud, S. (1915b). *Instincts and their vicissitudes*. S.E. XIV.
Freud, S. (1920). *Beyond the pleasure principle*. S.E. XVIII.
Freud, S. (1930). *Civilization and its discontents*. S.E. XXI.
Freud, S. (1940). *An outline of psycho-analysis*. S.E. XXIII.
Goozen, V., Stephanie, W., Victor, M., Endert, E., Helmond, F. A., de Poll, V., & Nanne, E. (1997). Psychoendocrinological assessment of the menstrual cycle: The relationship between hormones, sexuality, and mood. *Archives of Sexual Behavior, 26*, 359–382.
Gunderson, J. (1996). The borderline patient's intolerance of aloneness: Insecure attachments and therapist availability. *The American Journal of Psychiatry, 153*, 752–758.
Hazan, C., & Shaver, P. R. (1987). Romantic love conceptualized as an attachment process. *Journal of Personality and Social Psychology, 52*, 511–524.
Insel, T. R. (2003). Is social attachment an addictive disorder? *Physiology and Behavior, 79*, 351–357.
Insel, T. R., Winslow, J. T., Wang, Z., & Young, L. J. (1998). Oxytocin, vasopressin, and the neuroendocrine basis of pair bond formation. *Advances in Experimental Medical Biology, 449*, 215–224.
Karama, S., Lecours, A. R., Leroux, J.-M., Bourgouin, P., Beaudoin, G., Joubert, S., & Beauregard, M. (2002). Areas of brain activation in males and females during viewing of erotic film excerpts. *Human Brain Mapping, 16*, 1–13.
Kernberg, O. F. (1995). *Love relationships: Normality and pathology*. New Haven: Yale University Press.
Meyers, S. A., & Berscheid, E. (1997). The language of love: The difference a preposition makes. *Personality and Social Psychology Bulletin, 23*, 347–362.
Mikulincer, M. (1998). Adult attachment style and affect regulation: Strategic variations in self-appraisals. *Journal of Personality and Social Psychology, 75*, 420–435.

Mikulincer, M., & Shaver, P. R. (2003). The attachment behavioral system in adulthood: Activation, psychodynamics, and interpersonal processes. In M. P. Zanna (Ed.), *Advances in experimental social psychology* (Vol. 35, pp. 53–152). New York: Academic.

Mikulincer, M., & Shaver, P. R. (2007). Psychodynamics of attachment and sexuality. In D. Diamond, S. J. Blatt, & J. D. Lichtenberg (Eds.), *Attachment and sexuality* (pp. 51–78). New York: Analytic Press.

Mitchell, S. A. (1997). Psychoanalysis and the degradation of romance. *Psychoanalytic Dialogues, 7*, 23–41.

Mitchell, S. A., & Black, M. J. (1995). *Freud and beyond.* New York: Basic Books.

Panksepp, J. (1981). Hypothalamic integration of behavior: Rewards, punishments, and related psychobiological process. In P. J. Morgane & J. Panksepp (Eds.), *Handbook of the hypothalamus: Vol. 3, part A. Behavioral studies of the hypothalamus* (pp. 289–487). New York: Marcel Dekker.

Panksepp, J. (1998). *Affective neuroscience: The foundations of human and animal emotions.* New York: Oxford University Press.

Panksepp, J. (2003). At the interface between the affective, behavioral and cognitive neurosciences: Decoding the emotional feelings of the brain. *Brain and Cognition, 52*, 4–14.

Panksepp, J. (2005). On the embodied neural nature of core emotional affects. *Journal of Consciousness Studies, 12*, 161–187.

Panksepp, J. (2006). Emotional endophenotypes in evolutionary psychiatry. *Progress in Neuro-Psychopharmacology and Biological Psychiatry, 30*, 774–784.

Panksepp, J., & Biven, L. (2012). *The archaeology of mind: Neuroevolutionary origins of human emotion.* New York: Norton.

Panksepp, J., & Watt, D. (2011). Why does depression hurt? Ancestral primary-process separation-distress (PANIC) and diminished brain reward (SEEKING) processes in the genesis of depressive affect. *Psychiatry, 74*, 5–14.

Pfaff, D. W. (1999). *Drive: Neurobiological and molecular mechanisms of sexual motivation.* Cambridge, MA: MIT Press.

Pfaff, D. W., Martin, E., & Kow, L.-M. (2007). Generalized brain arousal mechanisms contributing to libido. *Neuro-Psychoanalysis, 9*, 173–181.

Schmidt-Hellerau, C. (2001). *Life drive & death drive – Libido & lethe: A formalized consistent model of psychoanalytic drive and structure theory.* New York: Other Press.

Solms, M., & Nersessian, E. (1999). Freud's theory of affect: Questions for neuroscience. *Neuro-Psychoanalysis, 1*, 5–14.

Solms, M., & Turnbull, O. (2002). *The brain and the inner world: An introduction to the neuroscience of subjective experience.* London: Karnac.

Stanley, B., & Siever, L. J. (2010). The interpersonal dimension of borderline personality disorder: Toward a neuropeptide model. *The American Journal of Psychiatry, 167*, 24–39.

Taylor, S. E., Klein, L. C., Lewis, B. P., Gruenewald, T. L., Gurung, R. A. R., & Updegraff, J. A. (2000). Biobehavioral responses to stress in females: Tend-and-befriend, not fight-or-flight. *Psychological Review, 107*, 411–429.

Tiihonen, J., Kuikka, J., Kupila, J., Partanen, K., Vainio, P., Airaksinen, J., Eronen, M., Hallikainen, T., Paanila, J., & Kinnunen, I. (1994). Increase in cerebral blood flow of right prefrontal cortex in man during orgasm. *Neuroscience Letters, 170*, 241–243.

Yovell, Y. (2008). Is there a drive to love? *Neuro-Psychoanalysis, 10*, 117–188.

Chapter 9
The Island of Drive: Representations, Somatic States and the Origin of Drive

Pierre J. Magistretti and François Ansermet

Abstract Freud defined the drive as "a concept on the frontier between the mental and the somatic". Today this view that was based on clinical observations interpreted within the psychoanalytical framework, can be revisited in light of the current neuroscientific notions of neuronal plasticity and somatic states. Indeed, through the mechanisms of plasticity experience leaves a trace that forms the neural basis of a representation of the experience. Such a representation R is associated with a somatic state S in the sense taken from the "somatic marker" model of Damasio. Thus, the internal reality of the subject, particularly the unconscious one, is constituted by such connected R's and S's. In the model discussed here, the posterior insula represents the primary interoceptive cortex where information about somatic states S converges, while in the anterior insula the connection between R and S can take place and establish a neurobiological correlate for the notion of drive. The authors posit that the re-representations of S associated with R in the anterior insula may correspond to the 'Vorstellungsrepräsentanz' postulated by Freud. They further propose that the tension between R and S, established in the anterior insula, is discharged according to the notion of drive through the motor arm of the limbic system, namely the anterior cingulate cortex which is heavily connected with the anterior insula.

Keywords Insula • Anterior cingulate cortex • Drive • Somatic states • Vorstellungsrepräsentanz • Freud

---

A previous version of this article was published in *Swiss Archives of Neurology and Psychiatry* 163 (08), 2012.

P.J. Magistretti (✉)
Laboratory of Neurogenetics and Cellular Dynamics, École Polytechnique Fédérale de Lausanne, EPFL-SV-BMI, SV2513, Station 19, CH-1015 Lausanne, Switzerland
e-mail: pierre.magistretti@epfl.ch

F. Ansermet
Department of Psychiatry, Université de Genève, HUG/SPEA,
Rue Verte 2, CH-1205 Genève, Switzerland
e-mail: francois.ansermet@unige.ch

© Springer International Publishing Switzerland 2016
S. Weigel, G. Scharbert (eds.), *A Neuro-Psychoanalytical Dialogue for Bridging Freud and the Neurosciences*, DOI 10.1007/978-3-319-17605-5_9

In *Instincts and their Vicissitudes/Triebe und Triebschicksale* (1915a), Freud defines *drive* as "a concept on the frontier between the mental and the somatic," more specifically as "the psychical representative of the stimuli originating from within the organism" (S.E. XIV, 121–122). According to psychoanalytical theory and in keeping with Freud's definition, drives play a role in the actions enacted by a subject. The drive that underlies an act by the subject originates from within the organism and, therefore, the subject's motivation for an action remains unknown and often surprising. This article contrasts the origin of an action with the decision-making process that belongs predominantly to the domain of consciousness (Fig. 9.1).

Through the mechanisms of *plasticity* experience leaves traces in neuronal networks that result in the establishment of an internal reality that is made both of conscious and unconscious representations. The latter engages a particular form of plasticity, namely the process of reconsolidation and trace re-association, which establishes a kind of discontinuity between experience and traces. The unconscious results from such discontinuities (Ansermet and Magistretti 2007a) (Fig. 9.2).

As William James formulated it years ago and more recently reelaborated by Antonio Damasio, somatic states play a central role in emotions and in the decision process. We propose that for each experience that produces a representation R through the mechanisms of plasticity, a representation of the somatic state S is also associated. The representations that constitute an internal reality, whether it be conscious or unconscious, are therefore associated with somatic states, thus establishing a strong association between R and S (ibid.; Ansermet and Magistretti 2010) (Fig. 9.3).

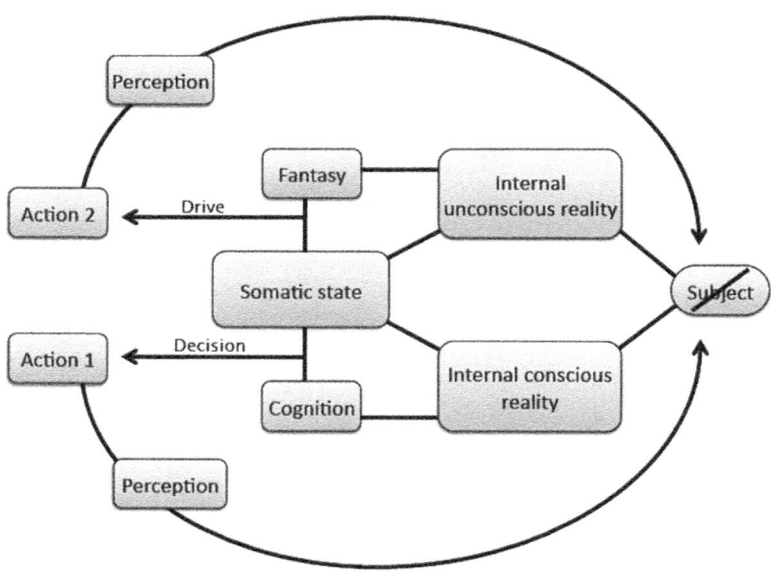

Fig. 9.1

9 The Island of Drive: Representations, Somatic States and the Origin of Drive

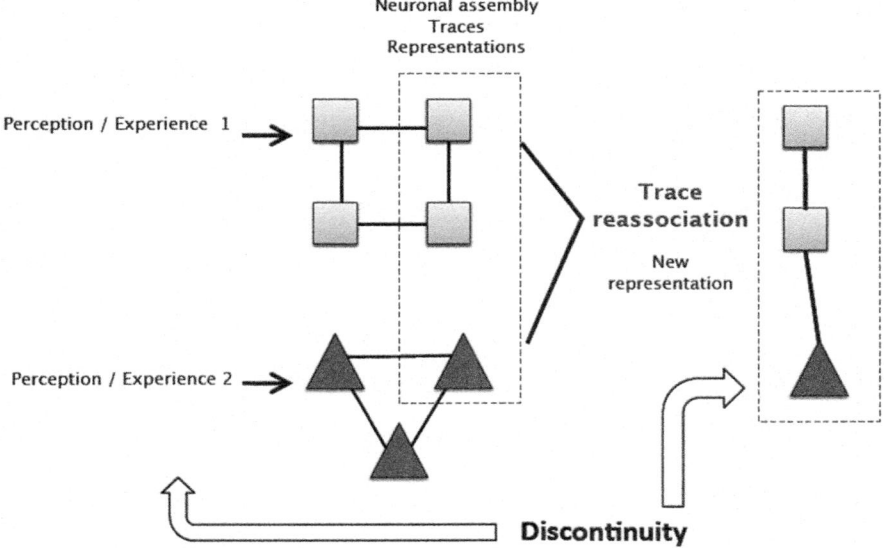

Fig. 9.2

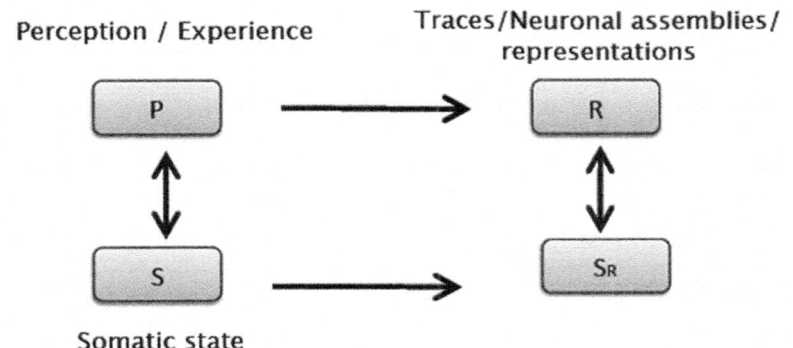

P : Perception/experience
R : Representation of P (eg neuronal assembly)
S : Somatic state
S_R : Representation of S ((eg neuronal assembly)

Fig. 9.3

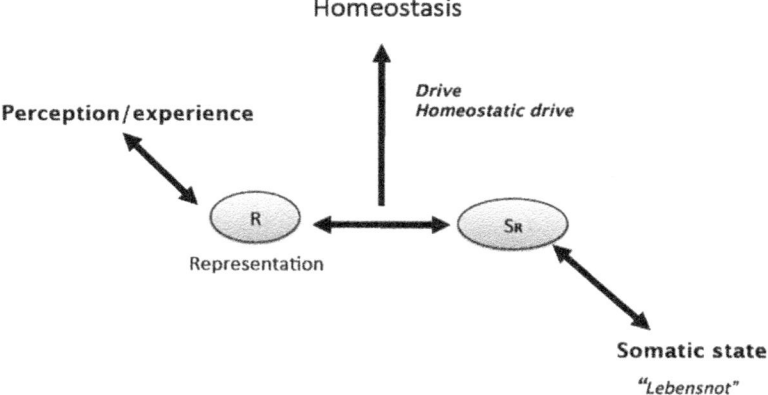

Fig. 9.4

According to this model, a drive is the unconscious counterpart to the decision process (Fig. 9.1). Indeed, Damasio's somatic markers theory (Damasio 1996) proposes that decisions rely on the anticipation of the somatic state in which the act produced by the decision will put the subject. Drive can be viewed as an "unconscious decision process" that emanates from the association of a somatic state and an unconscious representation or fantasy (Ansermet and Magistretti 2007b). The acts that derive from either decision or drive will be perceived by the subject and contribute to its own becoming (Fig. 9.1).

One can postulate that any S associated with an R potentially represents a deviation from the homeostatic state (Fig. 9.4). Thus, a pressure to reestablish the homeostatic state emerges, as with any physiological process and as proposed by the inertia principle of the psychoanalytical theory (Arminjon et al. 2010). However, because the somatic state S is associated with a representation R, the reestablishment of homeostasis will be constrained by the content of the representation R and therefore engage an action related to it. Consequently, the release of the act related to the drive depends on the nature of the representation R that is associated with the somatic state S (Fig. 9.4).

9.1 Sensing the Somatic States: The Interoceptive System

What is meant by the term somatic state? Which neuronal circuits mediate their perception? All organisms are exposed to stimuli that originate both from the external world but also from the interior of the body as Freud has already so beautifully

9 The Island of Drive: Representations, Somatic States and the Origin of Drive

described in his *Note on the Mystic Writing Pad* (1925). Exteroceptive sensory systems such as vision, hearing, olfaction and touch detect stimuli that originate from the external world. In contrast, the interoceptive system provides the means to detect the general state of the body, or the *Gemeingefühl* as it is called in German research literature. This system informs the brain about the state of viscera, glands and smooth muscles.

While the physiology of the exteroceptive system has been the object of intensive investigation over the last century, only recently attention has been given to the physiology of the interoceptive system, in particular thanks to the pioneering work of Bud Craig (2003, 2009). The interoceptive system is organized so that fibers enter the posterior horn of the spinal cord and progress towards higher brain centers (Fig. 9.5). These fibers have a small diameter and are only partially myelinated or

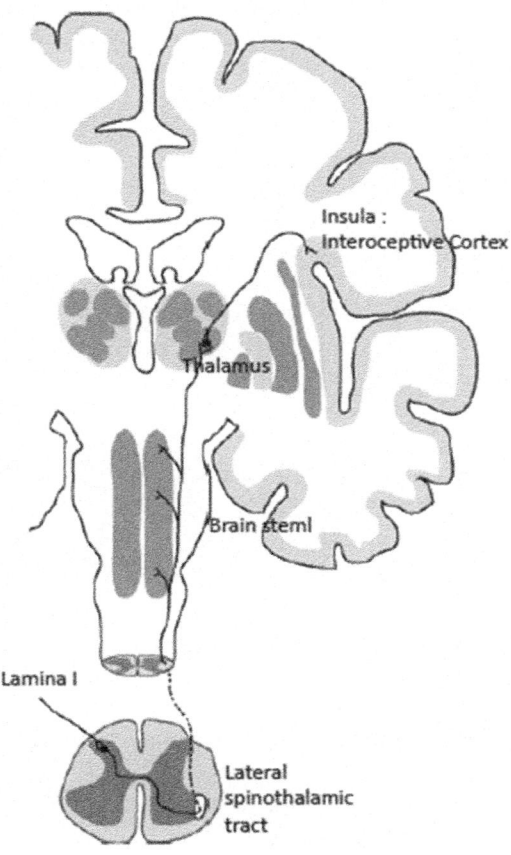

Fig. 9.5

even unmyelinated; hence their conduction velocity is relatively slow, on the order of one meter per second. These fibers are similar to the A Delta and C fibers that transmit pain sensation (nociception). After a first synapse in the posterior horn, the interoceptive fibers cross to the other side of the spinal cord and ascend towards brainstem nuclei, and in humans they progress to specific thalamic nuclei (Fig. 9.5). As noted by Damasio (2010), the processing of this interoceptive information already occurs in the brainstem.

From the thalamus, these fibers project to a specific cortical area, the insula, which can be considered the primary cortex for the interoceptive system. The insula, a kind of "island," is located at the interior aspect of the frontal lobes and is anatomically and functionally divided into an anterior and a posterior division. Interoceptive information is conducted through this ascending system terminating in the posterior insula. The interoceptive system shares some similarities with the somatosensory system that projects from the skin to a primary sensory area, specifically in the parietal cortex. Along with the automatic nervous system, the foremost function of the interoceptive system is to contribute to the maintenance of bodily homeostasis. The balance between the two divisions of the autonomic nervous system, namely the sympathetic and the parasympathetic divisions, which exert opposite effects, ensures the maintenance of homeostasis and the physiological adjustment of the vital functions of the organism, thus contributing to the maintenance of an organism's integrity. These effector systems that originate in brainstem nuclei and in the intermediolateral horn of the spinal cord represent the motor branch of a reflex loop. The sensory arm of this loop is in turn represented by the interoceptive afferents (Fig. 9.6).

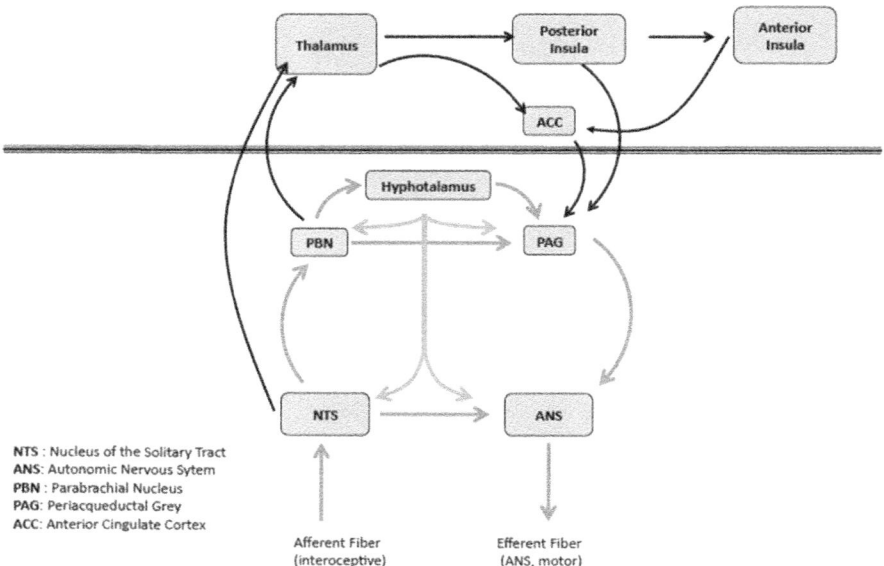

Fig. 9.6

This sensory-motor loop allows for the execution of fine physiological adjustments that contribute to the maintenance of bodily homeostasis. Functional connections between the sensory (interoceptive) and motor (autonomic nervous system) arms of this sensory-motor loop occur at different levels of the nervous system, notably at the spinal cord and brainstem levels. Other reciprocal connections exist between nuclei in the brainstem and the hypothalamus, contributing to the overall maintenance of bodily homeostasis. These connections rely upon mechanisms that can be characterized as being "automatic" or as a "reflex" between the sensory and motor arm of the homeostatic system. These reflex relationships clearly exist at lower levels of the nervous systems, for example in the spinal cord, the brainstem and the hypothalamus and do not engage with higher levels of the nervous system (Fig. 9.6).

However, in humans the maintenance of bodily homeostasis not only engages these reflex regulations but also involves more complex behaviors that mobilize areas of the brain involved with higher brain functions and, in particular, with executive functions. Even simple behaviors such as finding food or the search for a sexual partner imply an interplay between motivation, anticipation of pleasure/displeasure and reward, all of which are based on a process such as decision making on the conscious level, and at the same time they are based on drive on the unconscious level.

9.2 Beyond the Automatic Regulation of Homeostasis: Humans vs. Animals

In humans the homeostatic regulation mediated by the interoceptive system goes beyond the reflex level at the spinal cord, brainstem and hypothalamic levels. Indeed, the interoceptive system establishes synaptic relays in the thalamus and terminates in the posterior insula (Fig. 9.6). A representation of the bodily states is established in the posterior and dorsal insula, which is itself strongly connected with other brain regions involved in motivated behaviors. These regions include, for example, the anterior cingulate cortex or the nucleus accumbens, the latter being involved in reward mechanisms. Thus, it appears that the extension of the interoceptive system beyond the brainstem is a characteristic of primates, in particular humans. It is therefore important to underline this fundamental difference between humans and other species.

In lower species homeostasis is re-established as a reflex without mentation. In contrast, the information originating from the body in humans constitutes a primary representation in the posterior insula, which is then associated with other representations in secondary re-representations. This results in a mode of homeostatic regulation that involves considerably more freedom by escaping automatic and reflex regulation. The primary representation of the bodily states occurs in the posterior insula; this primary representation then integrates information originating from other sensory areas at the level of the anterior insula (Fig. 9.7). This process results in a re-representation of the bodily states (Craig 2009). Thus, at a given point in time the primary representation of the bodily states would be contextualized with

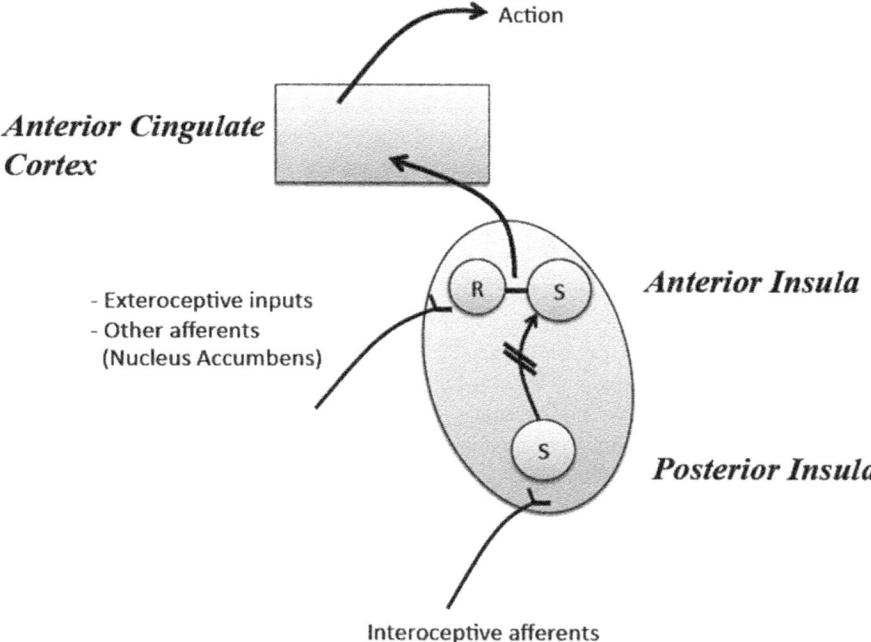

Fig. 9.7

information originating from other sensory areas. The information originating from the outside world through the exteroceptive systems would be integrated with information originating from the body and vehiculated to the anterior insula (Fig. 9.7).

This type of integration is similar to what occurs in exteroceptive systems, where the primary representations of the external stimuli occur in primary sensory areas; these primary representations are then integrated with other information at the level of secondary sensory areas, which enables the contextualization of the primary stimulus and its significance.

9.3 Are the Re-representations in the Anterior Insula the Neurobiological Equivalent to the Freudian Vorstellungsrepräsentanz?[1]

This notion of re-representation of the somatic states is also a key concept in psychoanalytical theory. Freud proposes the existence of a first representation of living matter, which would, in fact, correspond to the physiological state of the body; in

---

[1] The English translation of *Vorstellungsrepräsentanz* is "ideational representative" (S.E. XIV, 177).

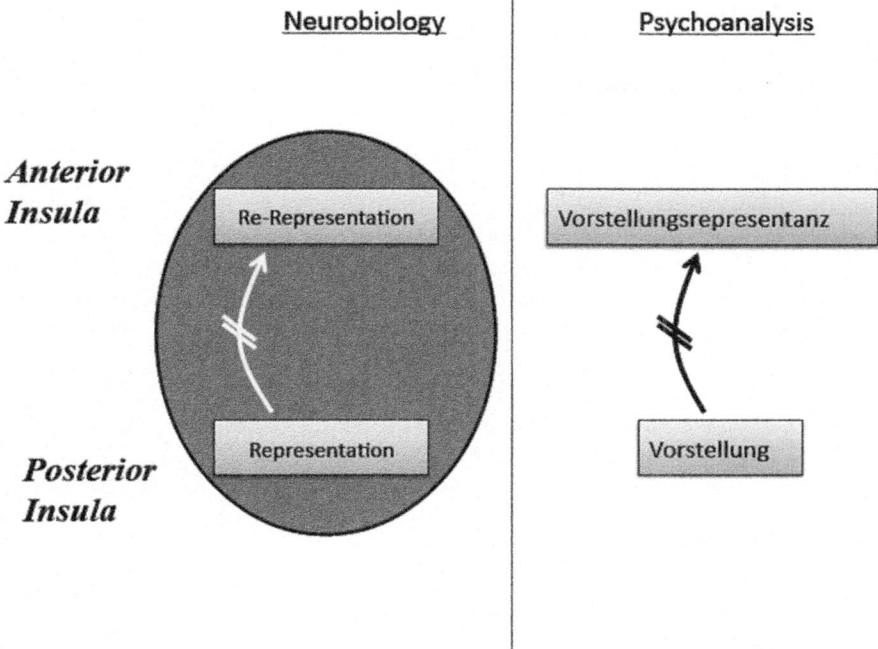

Fig. 9.8

his work, in *The Unconscious/Das Unbewußte* (1915b), on the concept of unconscious he defines this as *Vorstellung*. This *Vorstellung* would then be the object of a second representation for which he proposes the term *Vorstellungsrepräsentanz*, that is to say, the representative of representation. In Freud's view, the *Vorstellungsrepräsentanz* is a kind of ambassador, as Jacques Lacan (1979, 220) puts it, who represents living matter much like an ambassador represents the interests of his country in its interactions with foreign nations (Fig. 9.8).

The biological data on the interoceptive system support the existence of a re-representation in the anterior insula of the somatic state that was initially represented in the posterior insula. In the case of the anterior insula, the "foreign countries" would in fact be representations of other exteroceptive sensory modalities that provide a context for and associate with the representative of the somatic state detected by the interoceptive pathways. The anterior insula would therefore be a site of integration between the representation R of experience and the representations of the somatic states S (Fig. 9.7). Thus, the anterior insula is situated at a level where representations of the external world and their representations R become associated with representations of the somatic states S. One should not view this association that occurs in the anterior insula as a strict localization but rather as the crossroads of a dynamic process.

9.4 The Motor Aspect of Drive

In psychoanalytical theory the fundamental destiny of the drive produced by the tension between R and S is its discharge, which can only be achieved through the execution of an act, which implies a motor aspect to the drive (Fig. 9.4). Functional imaging studies have shown that the anterior insula is often co-activated with the anterior cingulate cortex. This observation is striking because the anterior cingulate cortex is strongly implicated in motor acts, in executive functions. This region of the brain plays a central role in what is known in neuropsychological theories as the motivation of the action. Accordingly, neuroanatomical as well as functional studies indicate a strong connection between the anterior insula and the anterior cingulate cortex. In fact these two regions are considered the sensory arm (anterior insula) and the motor arm (anterior cingulate cortex) of the limbic system. The anterior cingulate cortex also receives information concerning the bodily states through the interoceptive system (Fig. 9.6). Furthermore, the anterior insula and the anterior cingulate cortex are connected by a particular type of neuron that is found exclusively in humans, namely the Von Economo neurons. It is interesting to note that in frontotemporal dementia, a neurodegenerative disease which primarily affects frontal regions and Von Economo neurons in particular, patients show an emotional indifference and apathy that result in an inability to initiate actions.

This connectivity between the anterior insula and the anterior cingulate cortex would imply that the neuronal assemblies, which code the representations R and the associated somatic state S at the level of the anterior insula, can simultaneously activate neurons in the anterior cingulate cortex in order to initiate an action (Fig. 9.7). Thus, from a psychoanalytical point of view, one can postulate that this kind of connectivity would participate in the process of discharging a drive and its contribution to maintaining homeostasis.

9.5 Drive and Self-Consciousness

We have proposed a parallel between drives and decision making, the former emanating from unconscious representations or fantasies, the latter from conscious ones (Fig. 9.1). Freudian drives imply a discharge of excess living matter (*Lebensnot*, gr. *ananke*, Fig. 9.4) the latter described by Freud as a "natural law" and a main propulsion of the cultural development. While this is associated with the somatic state S through the mental content embedded in the representation R, one can nevertheless restate that the dynamic association between R and S occurs at the level of the anterior insula and that the discharge by the action occurs at the level of the anterior cingulate cortex and does so, at least partially, through the connections regulated by Von Economo neurons. Since a drive is discharged through an action, the subject perceives this action via exteroceptive systems such as vision or hearing (Fig. 9.1). While perceiving this act the subject is informed of the external manifestation of the somatic state at the origin of the drive.

In other words, and quite surprisingly, our internal bodily states manifest themselves with a slight temporal delay through the drive, therefore revealing our somatic state through the perception of the act. The subject is then divided by the act (Fig. 9.1).

Extending this viewpoint also has implications for self-consciousness. Perception of the somatic states at the level of the posterior insula through the activity of the interoceptive system, but especially its re-representation in the anterior insula, can contribute to the concept of self-consciousness (Craig 2009). However, the perception of the subject's action produced by the drive introduces a more dynamic aspect into the emergence of self-consciousness. Self-consciousness may also receive the impact of drives. Thus, self-consciousness not only appears to be related to cognitive processes in which interoception and the perception of the body play a key role. Self-consciousness also integrates aspects that include the dimension of drives and therefore indirectly dimensions of the unconscious.

References

Ansermet, F., & Magistretti, P. (2007a). *Biology of freedom: Neural plasticity, experience, and the unconscious*. New York: Other Press.
Ansermet, F., & Magistretti, P. (2007b). An unexpected phone call: How drives originate and what becomes of them. In F. Ansermet & P. Magistretti (Eds.), *Biology of freedom* (pp. 133–146). New York: Other Press.
Ansermet, F., & Magistretti, P. (2010). *Les énigmes du plaisir*. Paris: Odile Jacob.
Arminjon, M., Ansermet, F., & Magistretti, P. (2010). The homeostatic psyche: Freudian theory and somatic markers. *Journal of Physiology, 104*, 272–278.
Craig, A. D. "Bud" (2003) Interoception: The sense of the physiological condition of the body. *Current Opinion in Neurobiology, 13*(4), 500–505.
Craig, A. D. "Bud" (2009). How do you feel – now? The anterior insula and human awareness. *Nature Reviews Neuroscience, 10*(1), 59–70.
Damasio, A. R. (1994). *Decartes' error: Emotion, reason, and the human brain*. New York: Putnam.
Damasio, A. R. (1996). The somatic marker hypothesis and the possible functions of the prefrontal cortex. *Philosophical Transactions of the Royal Society of London. Series B, Biological Sciences, 351*(1346), 1313–1420.
Damasio, A. R. (2010). *Self comes to mind: Constructing the conscious brain*. New York: Pantheon.
Freud, S. (1915a). *Instincts and their Vicissitudes*. S.E. XIV.
Freud, S. (1915b). *The unconscious*. S.E. XIV.
Freud, S. (1925). *A note on the mystic writing pad*. S.E. XIX.
Lacan, J. (1979). *The four fundamental concepts of psychoanalysis*. (Jacques-Alain Miller, Ed., Alan Sheridan, Trans.). London: Penguin Books. (Original work published 1964)

Part V
Concerns of Psychoanalytical Theory

Chapter 10
Couch Potato: Some Remarks Concerning the Body of Psychoanalysis

Ulrike Kadi

Abstract Kadi indicates a shift from the former twofold understanding of psychoanalysis as consisting of therapy and research towards a threefold arrangement that includes therapy, research and the production of a metapsychology. This shift along with the technological progress in the field of neuroscience creates a potentially productive conflict for a neuroscientific reformulation of psychoanalysis. The author identifies the human body in its various conceptions as being at the core of this conflict. The body is seen alternately as a Lacanian mirror image, as an organless body bordering between the biological and the psychical, and as a medical body constituted by colorful but decidedly fantastic images can be developed through state-of-the-art imaging technology. Over and against these conflicting tendencies, Kadi calls for psychoanalysis to attend to the relevance of the neurological body and consider its relation to other concepts of the human body, in order for psychoanalysis to remain true to its calling to make the unconscious conscious.

Keywords Mirror stage • Lacan • Imaging technology • Metapsychology • Body/body image/body schema • Organless body

10.1 The Body as Frame of Reference

The South American psychoanalyst Isidoro Berenstein distinguishes three forms of practice in psychoanalysis: therapy with the patient, research into the unconscious, and the production of metapsychology as a basis for further developing psychoanalytic theory (Berenstein 2001, 141). This is an unusual trisection. The division into three fields is particularly surprising if one is used to Sigmund Freud's insistence upon only two things: the cure and research. For Freud they form an inseparable bond in psychoanalysis. Here we must, therefore, ask if Berenstein's suggestion contains an essential revision of Freud's objectives for psychoanalysis.

U. Kadi (✉)
Universitätsklinik für Psychoanalyse und Psychotherapie, Medizinische Universität Wien,
Währinger Gürtel 18-20, A-1090 Wien, Austria
e-mail: Barbara.Kadi@meduniwien.ac.at

Freud writes in the postscript to *The Question of Lay Analysis/Die Frage der Laienanalyse* (1926), "In psycho-analysis there has existed from the very first an inseparable bond between cure and research. Knowledge brought therapeutic success. It was impossible to treat a patient without learning something new; it was impossible to gain fresh insight without perceiving its beneficent results" (S.E. XX, 256). In this same text, Freud reflects on the medical treatment of patients. He emphasizes that there should not be a close tie between medicine and psychoanalysis. "[P]sychoanalysis is not a specialized branch of medicine" (ibid., 252). And the fact that it was developed by a medical doctor does nothing to change this. Take electricity for example: it did not become a specialized branch of physiology just because the first observations leading to its discovery were made on nerve cells (cf. ibid.). Freud describes himself in this text as an unambitious physician who only worked as a doctor out of financial necessity. His teacher Ernst Brücke had even warned him to avoid a career that dealt with theoretical issues—for the simple reason that he just would not be able to afford it. In the postscript Freud does note that since psychoanalysis is applied in patient therapy, it is often viewed as a specialized branch of medicine. But he wanted to make sure that "therapy will not destroy the science" (ibid., 254). Nevertheless, he concedes that treatment of patients is absolutely necessary for psychoanalysis itself, since the research material is easier to find in the treatment of neurotics than in observations of normal behavior (cf. ibid., 253).

This train of thought should be understood with Freud's concern about lay analysis in mind, which was very controversial at the time. As a lay analyst, Theodor Reik was involved in a court battle; he was accused of charlatanism. By shifting the objectives of psychoanalysis as far away from medical treatment as possible, Freud seems to give the impression that lay analysis is not comparable to medical therapy. But beyond this particular case, Freud's text clearly points to his own albeit limited interest in medical therapy, or rather to the attempt to limit psychoanalysis's scope when it comes to questions of therapy.

Some 75 years after Freud's *The Question of Lay Analysis*, Berenstein moves away from the Freudian division of research and therapy by splitting psychoanalysis into three parts: therapy, research into the unconscious, and metapsychology that informs psychoanalytic theory. He also mentions therapy first, something that should not be overlooked, as it is a clear indication that Berenstein puts therapy at the forefront of psychoanalysis. This reflects a development within psychoanalysis, in which therapy has gained greater meaning than Freud ascribed to it. It especially pertains to developments in Great Britain, where matters concerning therapy and technique following Melanie Klein and the post-Klein school are strongly emphasized. In the U.S.A., therapy and the medical application of psychoanalysis were the main issues. As Eric Kandel puts it, during the first half of the twentieth century psychoanalysis in America owed its good reputation to its connection with psychiatry (Kandel 1999).

This close connection to medicine had an impact on psychoanalysis and its work. In addition to a greater sensitivity for matters concerning therapy techniques, this impact pertains to the object of the research. Medicine revolves around the body.

10 Couch Potato: Some Remarks Concerning the Body of Psychoanalysis

Special branches of medicine, such as psychiatry, maintain this focus on the body, even though they hold a marginal position vis-à-vis other branches of medicine. Freud (1926) on the other hand warns against somatic medicine, when he says that medical analysts have to learn to "resist the temptation to flirt with endocrinology and the autonomic nervous system, when what is needed is an apprehension of psychological facts with the help of a framework of psychological concepts" (S.E. XX, 257).

Other than the fact that Berenstein puts therapy first, there is something else of note here: he divides the concept of research mentioned by Freud into two realms—studying the unconscious and building a theory. We ought to interpret Berenstein's distinctions with regard to Freud as an explanation that does not go beyond Freud but rather tries to make his reasoning even more precise, for research naturally implies building theories. The specific thing about psychoanalysis as a field of study is its research object, namely the unconscious. Berenstein underlines the fact that the unconscious has to be accepted as the central focus of psychoanalysis.

Through Berenstein's division into three forms of practice in psychoanalysis and the shift away from Freud that this entails, we are made aware of a conflict that is also reflected in neuro-scientific discussions. It is a conflict that, viewed superficially, exists between two forms of understanding psychoanalysis: (1) psychoanalysis as a form of treatment of diseases, which seems to be of special interest in neuro-psychoanalysis versus (2) psychoanalysis as epistemic and an epistemically critical instrument in a series of other sciences, such as the study of literature, philosophy, pedagogy, fine arts, or political science, to name a few. Within these various sciences, psychoanalysis makes the matter of the unconscious a subject of discussion. This form of psychoanalysis does not work towards a technique applied during treatment. It does not focus on process research or outcome research but instead deals with the theory of psychoanalysis and its relation to other sciences in the humanities.

It becomes evident that the three forms of practice do not always go hand in hand in psychoanalysis. This field of conflict does not make it easy for neuro-scientific theories, which belong to the field of medically applied psychoanalysis. On the one hand, great hopes are set on a neuro-scientific reformulation of psychoanalysis, to such an extent that some people even speak of neuro-psychoanalysis as if it were something totally new. On the other hand, neuro-psychoanalytic deliberations have a hard time gaining entry into the canon of psychoanalysis. It is my thesis that *the body and its meaning make up the core of this conflict*. Neuro-psychoanalytic reasoning can only be taken as viable input for psychoanalytic thinking once this conflict has been worked through. In the following, I would like to contribute to the process of addressing the conflict by outlining a few different positions regarding the body in psychoanalysis.

I agree with André Green's premise that the discoveries of psychoanalysis can only be found on the couch and not in a laboratory (Green 2004, 40). A body lies on the couch at a psychoanalyst's office. It is the same body that can be medically treated and neurologically examined. It is also the same body that appears as an object of our individually and culturally loaded associations. And, finally, it is that very same body that can become a source of anxiety. These moments can only be addressed if we place experience at the foundation of our reasoning (Katz 2010). Experience has to be viewed as the basis for this reasoning and especially of psychoanalytical theory. Our lived body (*der Leib*) is something other than a thing in a room (Merleau-Ponty 1973, 5) because we have a special relationship to it. We grasp it as *our* body. We cannot separate ourselves from it. It is characterized by a constant closeness to us (Merleau-Ponty 1966, 116). It is our frame of reference. It supports us. We need it. And even if we wanted to we could never get rid of it. Its three dimensions, as a medical body, a culturally formed body, and a body of fears, allow the body—and with it also the body on the couch—to become the knot that ties together different fields of experience in psychoanalysis.

10.2 The Projected Body

Some time ago, a colleague of mine told me about the difficulties she was having with her 14-year-old son. He was a good-natured amicable guy, and she never had much trouble with him—at least she claimed they seldom fought. But now something seemed to be wrong with him. She phrased it in the following way: he was a couch potato. As a psychoanalyst accustomed to delving into concrete words, the very phrase made me think of a small spud on a large sofa (cf. also Brown et al. 2010), which was actually how my great big colleague spoke of her little son. Before my very eyes, the son mutated into a small vegetable appendix of his all powerful mother. A little potato in danger of being chewed out by his mom, chewed and swallowed. As I listened to the mother talk on, this impression changed. My colleague depicted her son, more and more, as a monster besieging and even almost threatening the whole family. He occupied the couch. He never left it. He would permanently lie in front of the TV and even get his sisters to bring food to him there. My view of the son changed. The little teeny-weeny spud on a big sofa turned into a huge potato, stuffing its face with potato chips; the potato was spilling over the sides of the sofa: It was a lazy and limp hanging mass. And it was just hanging out there like some alien—an eerie life form that contrasted sharply with the family members who shrunk to the size of dwarves in the presence of this giant.

Several things come to mind when we speak of the body. We connect the body with a form. Bodies often lead us onto the subject of food. And usually we regard the body in relation to other bodies. In her book *Bodies* (Orbach 2009) the psychoanalyst Susie Orbach describes an array of these contemporary points of reference. If there are doubts, bodies can be surgically operated into shape. They are self-destructively starved or overfed. And they serve to make more or less temporary contact with other bodies in situations with stronger or weaker sexual connotations. Orbach shifts the passivity of the body into the foreground, when she points to the following: "One is not born a body, one becomes one" (Orbach 2009, 171).

What does this mean? In a study entitled *Eating Behaviours and Attitudes Following Prolonged Exposure to Television among Ethnic Fijian Adolescent Girls* (Becker et al. 2002) the authors demonstrate that the intrusion of television in a

Fig. 10.1 Erik Porath: Couch potato

previously TV-free zone in the Fiji Islands led to the first incidence of an anorexic eating disorder among girls (ibid., 169). Idealized body forms can invade a media-naïve population. Body images transmitted by TV or in films deliver body standards that are then transformed into body norms. The often contradictory intermingling of individual socialization with cultural standards and norms conveyed by mass media molds the shape of today's bodies. Decisive for this transfer of meaning is the visual standard. Eyes, it seems, are easier to manipulate than ears. A radio station in a formerly radio-free zone would probably not lead to such a quick and intense change in body images.

Bodies and their images are not simply there and tangible, but they are projected, formed, culturally provoked and re-shaped. That also goes for couch potatoes, who stereotypically represent a certain type of body glued to the TV. Bodies develop based on psychical processes, through processes of recognition, individuation, and separation. As Freud writes in *The Ego and the Id* (1923), "[t]he ego is first and foremost a bodily ego; it is not merely a surface entity, but is itself the projection of a surface" (S.E. XIX, 26).

In psychoanalysis one's own body, therefore, plays a central role in a subject's development. The body image and the body schema, which the body develops in its first years of life, are the consequences of a child's complex early experiences with its first attachment figure. Mirror experiences with the body of the parent and with that other body, which cannot yet be adopted as one's own, need to be integrated. This leads to confusion, illusions, and delusions, even to psychotic intrusions because a body image that is still instable distorts the patient's connection to reality (Widmer 2007, 87). Later during development, paranoid experiences of doppelgangers might emerge that remind the subject of its own uncertain origin.

They in addition supply the material for individuation and separation that are necessary steps in the development of the subject.

The doppelganger motif—an oppressive experience of having one's body invaded, permeated or at least defined by the body of another—has been examined more closely by Otto Rank, another lay analyst from the early days of psychoanalysis. This motif contains some of the horror of the psychotic, a terror that goes beyond a single lifetime and is culturally determined. Rank devoted himself to how this terror has been repeatedly reshaped (cf. Dolar 1993, 125). Jacques Lacan's well-known mirror stage, which describes the genesis of the subject from the body image of the other, also picks up on the doppelganger motif and shifts the dialectic moment of conquering fear into the foreground. It is a dialectic interaction between the not-yet-subject, who develops into a subject with the help of the mirror, and its first attachment figure, who is now an agent in a formative social matrix. The Hegelian struggle between master and servant determines the course of this development. The child, thus, faces two possibilities. The struggle between the self and the other leads to only one question: Is it me or you? The body, which is involved in this game, presents itself as a sort of battleground. Initially deferred it returns again fragmented—deformed in the dreams of the subject, or distorted in artwork as in Hieronymus Bosch's paintings (cf. Lacan 1973, 67).

The realm of the doppelganger is the realm of the mirror. In neuro-physiological terms, we speak of mirror neurons. This concept is based on the observation that one person's neurons fire when another person is doing something. This concept of identification helps us understand how people learn by imitation. But it seems to be rather insufficient in describing the aggressive dialectic between a not-yet-subject and another subject, that inter alia is perceived as a threatening enemy during separation. The psychic struggle with the threatening other leads to visions of a fragmented body. The image of a coherent body form can be understood as an expression of integrating the conflicting experiences of a body in pieces (*le corps morcelé*) (cf. ibd.). Lacan emphasizes the instable basis of this ability to integrate when he attributes thousands of misleading characteristics to the image of one's own body and to the fact that the emerging subject clings to these images.

Our body image as well as our capacity for empathy is a mixture that depends on our own personal histories. People who react in an empathetic way are simultaneously dealing with previous threatening moments from their own affectivity, such as envy, hate and resentment, but in a socially acceptable way. They thereby repeat a gesture they have developed towards the image of the body in pieces: They oppose a coherent behavior to an experience of a disrupted condition. But negative feelings only seem to disappear. It is because of such masking effects that Lacan warns against "altruistic feelings" (Lacan 1973, 70).

Couch potatoes also embody the results of confrontations. A fragile "I" is confronted with external social demands. Couch potatoes do not participate and do not adhere to prescribed rules. They are cheaters. In fact, the term "coach potato" has come to mean just this in the realm of Geocaching. In this contemporary variation of a scavenger hunt, players locate places throughout the world where they come across several questions. The honest players take these questions from the places out in the world and answer them later on the Internet. A couch potato breaks the rules and does not go out and search in the field. Instead the couch potato player stays at home and surfs the Internet for the solution.

10.3 The Organless Body

The phrase of a potato on the coach evokes the image of not an animal or human but a vegetable body. A potato has brown skin and yellow insides. The inner part forms (as long as the potato is not attacked by pests or turned rotten for some reason) a homogenous structure. From a macroscopic perspective, and disregarding the function of its skin, we see a body without organs. It represents a further dimension that we might ascribe to the body. Bodies without mirrors and without knowledge about their innards are simply present: a thick mass without any further structure.

The organless body stems from the imagination of the poet Antonin Artaud. It does not belong to our everyday experience of bodies. The philosopher Gilles Deleuze and the psychoanalyst Félix Guattari draw on Artaud's imagination in their *Anti-Oedipus* (1974). An organless body is a body standing for an opposition towards our everyday experience of bodies. It has a series of facets, which are of interest to us here (cf. for the following also David-Ménard 2003, 79 f.): Artaud introduces it as a way to free the body from its automatisms. An organless body is a body that directs itself against an "organic and sexual totality" (ibid., 80). Deleuze and Guattari particularly like to speak of an organless body because it offers the opportunity of making the connection between madness and artistic production in terms that are not negative (ibid., 82 f.). The notion of an organless body became not so much a criticism of society, as was Artaud's intention, but a criticism of psychoanalysis itself. Their critique was, according to David-Ménard, justified with regard to a certain psychoanalytical ideology. Still, it remains a monstrous oversimplification (ibid., 85).

The organless body, we might say, represents a body in resistance, a body which has become resistance itself. It resists common ascriptions to the concept of the body. It calls to mind antibodies, which in turn remind us of the confrontation with infectious agents. With the idea of an organless body, we are faced with a body in a political as well as psychological sense—a body that cannot lead to a unified body on its own since it does not represent diversity; a lonely body since it does not find itself in the field of the other; a body forced to produce affects on its own from itself. But most importantly, the notion of an organless body provides a connection to the "drive," a term in Freudian psychoanalysis, which Marc Solms points out is gradually disappearing from psychoanalytic theory (cf. Solms and Turnbull 2002, 131). "The organless unified body is unproductive, sterile, misconceived, inedible… Death drive is its name" (Deleuze and Guattari 1974, 14).

As already mentioned, Deleuze and Guattari use the organless body as an instrument to criticize psychoanalysis. This might explain why the organless body did not arouse much interest within psychoanalysis itself. Slavoj Žižek points out that

Lacan connects the drive with one single organ and *not* with a whole impenetrable and unstructured body, as Deleuze and Guattari in their conception of the death drive do. Lacan calls this organ a *lamella*. It is something extra flat, something that moves like an amoeba. It can go anywhere, it has access to everything. It is an uncanny thing—according to Lacan, one should try to imagine it crawling across one's face (cf. Lacan 1978, 197). It is just as eerie as a couch potato, slumping over the sides of the sofa.

A lamella is a bodiless organ, an organ without body, like wandering body parts in dreams or in surrealistic films, or the Cheshire Cat's grin in *Alice in Wonderland*, which smiles on its own without the cat's body. Žižek connects the lamella with a Freudian partial object, a "bizarre organ that has magically gained autonomy and survived without the body of which it should have been an organ" (Žižek 2008, 12). This organ is to Lacan what the libido is to Freud. The lamella does not exist; instead, it insists and incorporates in this insistence a second characteristic, which Freud attributes to the drive, the persistent component of the drive, the death drive (cf. ibid., 13).

The drive, which we are referring to here is, in Freudian terms, something on the border between the biological and the psychical (cf. Ansermet and Magistretti 2005, 155 ff.). Therefore, the question whether to translate Freud's *Trieb* as "instinct" in English (as Strachey did regarding Freud's own ambiguity (Laplanche 2003) and as Solms will not do in the revised edition of Freud's complete work (cf. Solms 2012, 54)) has to do mostly with the two-sided position of the function of the driving force. All in all, Freud's drive incorporates something which is hard to imagine today: a dialectic interaction between an inborn instinctively structured, somatically transmitted impulse and a force continually redetermined and shaped by cultural guidelines. This contradictory dimension of Freud's *Trieb* is in danger of being lost if the question of the drive is mainly connected with a SEEKING-system and a LUST-system (Panksepp 1998) and with the activation and deactivation of "need detector mechanisms" (Solms and Turnbull 2002, 115 ff.). The potentially conflict-causing dimension of the drive that Freud foregrounded and that includes the intervention of the other (Laplanche 1988) gets occluded by these terms.

The organless body or the bodiless organs can be grasped as the phantasm of that same body that was worked over by the drive theory. Here we are dealing with an uncanny body, a human body which does not exclude its biological and animalistic aspect but in fact includes this irritating and dangerous side.

10.4 The Medical Body

No one wants to be a couch potato. The bodies of couch potatoes are not desirable, not only due to their appearance but also because they are expensive to maintain and present a higher risk of disease. Therefore great efforts must be made to transform these bodies. A vast amount of research over the past few years supports these findings: *The High Cost of Couch Potatoes* (Chenoweth and Pfohl 2000), *The Cost of*

Being Couch Potato (De Jong et al. 2003), *Getting Fit. Couch Potatoes, Arise!* (Kluger 2005). And: *Couch Potatoes to Jumping Beans: A Pilot Study of the Effect of Active Video Games on Physical Activity in Children* (Ni Mhurchu et al. 2008). Couch potatoes ought to disappear, or better yet never be allowed to take root in the first place. The research on couch potatoes is principally concerned with prevention. This is perfectly understandable from an economic perspective, as well as from the point of view of internal medicine or neurology. After all, due to their high cholesterol levels, the risk of heart attack for couch potatoes is clearly greater than for people who exercise regularly and eat a balanced diet.

These studies on coach potatoes address the body in medicine and in medical treatment. This is a body that has been available to us only in the past few centuries. It is a body that, as Lacan points out, owes its existence to "human unity, which that idiot Descartes had cut in two" (Lacan 1988, 73). It is a body which speaks out and of which one can speak. It is also a body that everyone keeps making pictures of. Just consider the diverse imaging technology processes from X-rays to magnetic resonance tomography. There are processes that set new goals based on the insistent nineteenth century wish to localize psychical potencies, by providing an inner map, where increasingly subtle degrees of structures could be depicted.

Lacan stresses the fact that it is impossible to retract Descartes' steps. Medicine's body is taken for granted in today's world. Medical professionals know the whole song and dance by now. They have seen patients who are obviously suffering but who do not grumble about their "sadness," "apathy," or "anguished souls." Instead they complain about their neurotransmitter deficits, and some even refer specifically to their serotonin deficiencies (cf. Ehrenberg 2004, 209 ff.). As a pictographic body, where individual sections of the brain are colorfully indicated or photographed in a true-to-life fashion, the medical body provides a wide and vast surface to project our fears and fantasies. In these pictures we find a body in which everything has its special order, a spatial order which Lacan calls "miraginaire" (Lacan 1988, 310). This term is a neologism, one of many that Lacan uses (cf. Bénabou et al. 2002). It is a contamination, a combination of the two French words *le miroir* (the mirror) and *imaginaire* (imaginary). It "was coined to stress the importance of mirror-stage identification with images or forms in the process of ego building" (Ragland-Sullivan 1986, 146). Seduced by body images the subject is deceived, a self-deception that Lacan describes in the mirror phase. The Cartesian body can be misled and the subject assumes that this body is its primary being.

In his *Psychopathology of Everyday Life* (1901), Freud clearly sets himself against post-Cartesian reductionism and adheres to the tenet that psychoanalysis does not primarily deal with physiological conditions but rather with psychological explanations (S.E. VI, 21). So we have to be careful, because "we would be making a serious mistake if we concluded that the associated mental illness itself can be described or understood from a physiological point of view" (cf. Solms 1995, 119). Admittedly, Freud's tenet is over one hundred years old. Some claim that if Freud lived in this day and age, his position towards correlations between physiological states and psychological structures would be different due to advances in neuro-physiological technology and knowledge (Solms 2000). But it cannot be overlooked that lots of methodological differences between psychoanalysis and neuroscience do oppose the establishment of sufficiently strong correlations. At the center of psychoanalytic theory is the couch, upon which, as likewise in society, a constant array of fears and conflicts are actually manifested during the attempt to alleviate these fears. As far as the body is concerned these fears pertain to our origins, our desires and our death. Sexuality is a central issue of these fears—concerning our origin the sexual intercourse of the parental figures makes up a focus of conflict which is renewed in experiences and phantasies of the primal scene. Our desire is shaped by enigmatic messages from the other (Laplanche 1992) which we have tried to decode with our infantile sexual theories. And our fear of death is connected to the sexually determined oedipal rivalry. The medical body itself can be understood as an attempt to conquer fears of this sort and accompanies other efforts to come to terms with frightening experiences. Out of these diverse efforts other body images grow: mirror images, images of lamella, and organless bodies. These different body images influence our cultural self-understanding. They are the object of a conflict over which images provide the greatest protection against unclear dangers.

On the couch of psychoanalysis there is no room for a couch potato. A psychoanalyst's couch has no TV in front of it, and potato chips are seldom crunched there. But the couch of psychoanalysis is shaped by images, often by body images. Since these body images serve as a defense mechanism, they tend to become rigid and freeze. To work through them they have to be transformed into speech so that frozen images can start to flow. Different images of the body will arise, some of which have been depicted in this article. Bodies in front of a mirror stand alongside organless bodies and medicine's bodies.

The couch of psychoanalysis is a kind of laboratory where, like under a microscope, cultural and individual associations with the body can be explored. It is also evident on this couch that the body of neurophysiology has an especially high potential to cope with fears and anxieties. Psychoanalysis cannot circumvent this body and its images. Whether psychoanalysis itself will behave like a couch potato and just sit back and watch the pictures of the neurologically described body like images on a television, or whether it will use the neurophysiological body in order to picture something in relation to other images that help make the unconscious conscious, those are questions that psychoanalysts will have to deliberate over in the years to come.

References

Ansermet, F., & Magistretti, P. (2005). *Die Individualität des Gehirns*. Frankfurt: Suhrkamp.
Becker, A. E., Burwell, R. A., Gilman, S. E., Herzog, D. B., & Hamburg, P. (2002). Eating behaviours and attitudes following prolonged exposure to television among ethnic Fijian adolescent girls. *British Journal of Psychiatry, 180*, 509–514.
Bénabou, M., Cornaz, L., de Liège, D., & Pélissier, Y. (2002). *789 Néologismes de Jacques Lacan*. Paris: Epel.
Berenstein, I. (2001). The link and the other. *International Journal of Psychoanalysis, 82*, 141–149.

Brown, J. E., Broom, D. H., Nicholson, J. M., & Bittman, M. (2010). Do working mothers raise couch potato kids? Maternal employment and children's lifestyle behaviours and weight in early childhood. *Social Science & Medicine, 70*(11), 1816–1824.
Chenoweth, D., & Pfohl, S. (2000). The high cost of couch potatoes. *Business and Health, 18*(1), 20–22.
David-Ménard, M. (2003). Was tun mit dem organlosen Körper? In É. Alliez & E. von Samsonow (Eds.), *Biographien des organlosen Körpers* (pp. 78–94). Wien: Turia+Kant.
De Jong, G., Sheppard, L., Lieber, M., & Chenoweth, D. (2003). The cost of being couch potato. *Michigan Health & Hospital, 39*(4), 24–27.
Deleuze, G., & Guattari, F. (1974). *Anti-Oedipus*. Frankfurt: Suhrkamp.
Dolar, M. (1993). Otto Rank und der Doppelgänger (Preface to Otto Rank: *Der Doppelgänger*, pp. 119–129). Wien: Turia+Kant.
Ehrenberg, A. (2004). *Das erschöpfte Selbst: Depression und Gesellschaft in der Gegenwart*. Frankfurt/New York: Campus Verlag.
Freud, S. (1901). *The psychopathology of everyday life*. S.E. VI.
Freud, S. (1923). *The ego and the Id*. S.E. XIX.
Freud, S. (1926). *The question of lay analysis*. S.E. XX.
Green, A. (2004). Pluralität der Wissenschaft und psychoanalytisches Denken. In M. Leuzinger-Bohleber et al. (Eds.), *Psychoanalyse als Profession und Wissenschaft* (pp. 33–48). Stuttgart: Kohlhammer.
Kandel, E. R. (1999). Biology and the future of psychoanalysis. *American Journal of Psychiatry, 156*, 505–524.
Katz, S. M. (2010). A holistic framework for psychoanalysis. *The Psychoanalytic Review, 97*, 107–135.
Kluger J. (2005, June 6). Getting fit. Couch potatoes, arise! *Time, 165*, 52–53.
Lacan, J. (1973). Das Spiegelstadium als Bildner der Ichfunktion, wie sie uns in der psychoanalytischen Erfahrung erscheint. In L. Jacques (Ed.), *Schriften I* (pp. 61–70). Olten: Walter Verlag. (Original work published 1966)
Lacan, J. (1978). *The seminar of Jacques Lacan. Book XI. The four fundamental concepts of psycho-analysis*. (A. Sheridan, Trans.). New York: Norton. (Original work published 1964)
Lacan, J. (1988). *The seminar of Jacques Lacan. Book II. The ego in Freud's theory and in the technique of psychoanalysis*. (S. Tomaselli, Trans.). New York: Norton. (Original work published 1954–1955).
Laplanche, J. (1988). *Die allgemeine Verführungstheorie und andere Aufsätze*. Tübingen: Edition Discord.
Laplanche, J. (1992). Interpretation between determinism and hermeneutics: A restatement of the problem. *International Journal of Psycho-Analysis, 73*, 429–445.
Laplanche, J. (2003). Trieb und Instinkt. *Forum der Psychoanalyse, 19*, 18–27.
Merleau-Ponty, M. (1966). *Phänomenologie der Wahrnehmung*. Berlin: Walter de Gruyter.
Merleau-Ponty, M. (1973). *Vorlesungen I*. Berlin: Walter de Gruyter.
Ni Mhurchu, C., Maddison, R., Jiang, Y., Jull, A., Prapavessis, H., & Rodgers, A. (2008). Couch potatoes to jumping beans: A pilot study of the effect of active video games on physical activity in children. *International Journal of Behavioral Nutrition and Physical Action*. doi:10.1186/1479-5868-5-8.
Orbach, S. (2009). *Bodies*. New York: Picador.
Panksepp, J. (1998). *Affective neuroscience: The foundations of human and animal emotions*. Oxford: Oxford University Press.
Ragland-Sullivan, E. (1986). *Jacques Lacan and the philosophy of psychoanalysis*. Chicago: University of Illinois Press.
Solms, M. (1995). Is the brain more real than the mind? *Psychoanalytic Psychotherapy, 9*, 107–120.
Solms, M. (2000). Freud, Luria and the clinical method. *Psychoanalysis and History, 2*, 76–109.

Solms, M. (2012). Are Freud's "erogenous zones" sources or objects of libidinal drive? *Neuropsychoanalysis, 14*, 53–56.

Solms, M., & Turnbull, O. (2002). *The brain and the inner world: An introduction to the neuroscience of subjective experience.* New York: Other Press.

Widmer, P. (2007). Die konstitutive Bedeutung des Körperbildes: Zur Begründung der Körperbildtherapie. In H. Lang, H. Faller, & M. Schowalter (Eds.), *Struktur, Persönlichkeit—Persönlichkeitsstörung* (pp. 87–104). Würzburg: Königshausen und Neumann.

Žižek, S. (2008). *Immer Ärger mit dem Realen. Troubles with the real.* Wien: Sonderzahl.

Chapter 11
"The Medulla Oblongata Is a Very Serious and Lovely Object." A Comparison of Neuroscientific and Psychoanalytical Theories

Edith Seifert

Abstract Seifert begins her article with the hypothesis that neuroscience can provide a theoretical framework for psychoanalysis, but her analysis ultimately shows this claim to be inadequate. Although Freud's dictum that man is "not even master in his own home" applies equally to the logic of the brain and central nervous system as it does to the logic of the psyche/the unconscious, there is still a decisive difference when it comes to the unconscious and Freud's often neglected topographical definition of it as a place of alterity, inaccessible by the conscious and perpetually alien. According to Seifert, we need to be more aware of the disconnect between body and soul and recognize that both psychoanalytic theorems and neuroscientific discoveries are rooted in language. This means that the laws of neuroscience and the laws of psychoanalysis are indeed not the same and that an epistemic leap across the gap between the empirical objectivity of neurons and a symbolic order of the psyche is necessary.

Keywords Linguistic neuroscience • Structural psychoanalysis • Benjamin Libet • E. Du Bois-Reymond • Lacan • Psycho-physical parallelism • J. M. Charcot

Like other disciplines of applied knowledge psychoanalysis is in constant need of alignment with neighboring disciplines. The various strands of neuroscience, the innovative power of which can hardly be denied, have been recommending themselves as such a point of reference for quite a while. Indeed, there appear to be many reasons to greatly appreciate neuroscience. Firstly, collaboration with the field of neuroscience holds the promise of curing illnesses such as autism, mental retardation, and cognitive disorders linked to Alzheimer's or Parkinson's disease. Secondly, neuroscientific research in combination with therapy prove the relevance of the psychoanalytic method. And finally, neuroscience might provide a theoretical

---

E. Seifert (✉)
Psychoanalytischer Salon Berlin, Laehr'scher Jagdweg 26, D-14167 Berlin, Germany
e-mail: e.seifert@kaleidoskopien.de

framework for psychoanalysis; it might, in other words, establish general laws and a level of abstraction that many psychoanalysts have been missing for long.

I would like to discuss this final point here. First of all, it touches on a debate that began long ago, a debate on theorizing psychoanalytical clinical practice centered on Freud's so-called metapsychology, which from the beginning was considered epistemically controversial and in need of improvement. In the 100-year history of psychoanalysis, discussions about the status of metapsychology have not ceased, returning time and again from various angles. Jürgen Habermas (1977, 262–332) and Paul Ricoeur (1970) lead discussions on the topic in the 1960s for example, and in the 1980s Adolf Grünbaum (1988) reconstructed them; today, neuroscience has taken them up again and has apparently reached a conclusion in its favor—that is, in favor of the necessity now more than ever of either orienting psychoanalysis along scientific lines or bidding it adieu once and for all as a hermeneutic science in search of meaning.

The debate provides the background for this chapter. I could begin by praising neuroscience for its work on the definition of metapsychology. I could mention that neuroscience has elevated our thinking about the psyche from the purely notional and the imaginary and has linked it to processes that follow specific laws. In spite of whatever objections might be raised, these other aspects ought to be seen as positive gains for science. But I would like to be more precise about what these laws are considered to be in a neuroscientific context.

The insight that cognitive, mental, and psychical processes display regularities can be traced back largely to the research and discoveries of Benjamin Libet's *Mind Time* (2004). In the late 1960s, Libet's experiments on the link between the conscious will and brain functions revealed that conscious decisions are preceded temporally by specific processes, and that these processes control such decisions. At that point, it became a generally accepted opinion that decisions and acts of will become conscious only at the end of a practically concluded process and not at the beginning—as was previously assumed. Libet's discovery, which predated Antonio Damasio's, demonstrates that "our consciousness is hopelessly delayed, in other words, perceptions of the consciousness represent ex post effects of neuronal processes that have already run their course" (Damasio 1994). Inasmuch as these earlier processes—and this is the decisive point—are also considered to be pre-personal processes (Pauen 2001, 293), we find ourselves facing a dimension that justifiably piques the interest of a psychoanalyst.

Now, hypothetically, if we momentarily ignore the fact and simply try out the idea that such pre-personal neurological processes are of course identified with the brain system, or with neuronal activity that occurs in the supplementary motoric area, we might indeed be tempted to consider this to be a confirmation of an important assumption on the part of psychoanalysis, albeit structural psychoanalysis. After all, according to structural psychoanalysis, the processes of human psychology are not simply limited to individual people's ideas but rather simultaneously run their course on a plane "beyond" the imaginary. Freud's dictum, in his *Introductory Lectures on Psychoanalysis/Vorlesungen zur Einführung in die Psychoanalyse* (1917), that man is "not even master in his own home" (S.E. XV, 285) applies eminently in this context. It means nothing less than that the individual is integrated in an order dominated by an alien logic that does not affect him or her as a person.

But this is where the intellectual game ends, because ultimately, everything depends on what this logic is called and whether it is accurately described by the logic of the central nervous system (CNS) and the brain, or whether it might not actually be expressed more appropriately by the logic of the unconscious when it comes to psychical processes like wanting, loving and desiring. I suggest that we now turn our attention to two alternative orders.

First, there is the order of the cerebral nervous system, the brain. According to Emil Du Bois-Reymond (1887) it is guided by the two principles of physicalism that assert that a system can be deemed physical if it obeys (1) the principle of causal unity and (2) the principle of physical determination (see Pauen 2001, 28 f.). Applied to the physical definition of the brain, I would describe it as follows: First, the brain is a complex, organized system that actively processes information defined purely along physical lines. Second, one of its special features is its recursive way of functioning, which requires that each element of the system be involved in producing the other elements. Another remarkable fact about the brain system is that all operations necessary to the system are generated by the interaction of the elements themselves. This principle, the closed nature of its operations, also guarantees that a system like the brain can function entirely autonomously.

The aspect of the autonomy is of considerable importance when it comes to environmental dependency, also known as the *plasticity* of the brain, as Wolf Singer in his concept of the "cooperative cell" and others explained in the 1980s (Singer 1990). Singer discovered this plasticity of the brain in the course of his research on neuronal maturation processes related to the visual perception of highly developed mammals. His findings are common knowledge today. When examining young cats, he discovered that their perceptions were not simply genetically determined but also had to be learned. He found that this learning takes place by means of experiences in one's environment and that a lack of experience, or deprivation of experience, brings about irreversible damage and can halt development at a stage of immaturity. In scientific terminology, this momentous discovery meant that functions of the human brain, too, developed through interaction with one's surroundings. In other words, in addition to their genetic repertoire they are also epigenetic, that is that they depend on experience and are a matter of plasticity.

I would like to briefly mention that in the case of the brain, two behaviors that are usually considered opposites, namely environmental dependence and autonomy, do not necessarily contradict one another. This claim is particularly important to keep in mind when assessing the role of psychical aspects. With this very brief presentation of some basic principles of the cerebral system in place, let us now turn to the psychical system of psychoanalysis: the unconscious. It may surprise some readers to learn that, at first glance, there are some astounding parallels between the cerebral system and the structure of the psychical system. For example, following Jacques Lacan, it would be an epistemic fallacy were we to regard the psychical apparatus as something other than a surface that tends to function more in terms of physical elements and quantities rather than in terms of hermeneutically detectable ones.

Freud, in his early neurological period, provided the model for Lacan. In terms of the theory of science, Lacan's psychoanalysis (Lacan 2006) takes natural science to such great lengths that it even supports Norbert Wiener's concepts, according to which the behaviors of machines and living organisms display many similarities when it comes to control and communication. As if tailored to Libet's experiment, Lacan, in conjunction with Freud, explains that the conditions and senses of the ego are always tied to formal requirements, and he emphasizes that a human operation such as thinking takes place without involvement of the ego. To put this idea another way, it would basically be impossible to say that it is 'I' who is thinking in the moment of thinking. Elsewhere Lacan emphasizes that in the case of certain vicissitudes of drives, such as perversions, the subject takes on the position of an object, whereby his or her objectivity does not at all harm the sense of ego.

Many more examples could further support the impression that structural psychoanalysis indeed allows for an objectified subject, both in terms of ego and of psychical processes. Even the unconscious, which is generally considered the most essential and the most intimate part of the subject, seems to display general traits that are definitely analogous to machines in appearance. According to Lacan, it has all the features of a primitive, organized switch process, and its (binary) organization is similar to that of machines with internal accounting systems. Since these processual aspects are quite remarkable when found in the most intimate part of the subject, one might get the impression that structural psychoanalysis would welcome the connection to neuroscience. Yet, I would argue, perhaps in a simplified way, that the opposite is true: For the small but decisive differences emerge the moment when we look more closely at the notion of the unconscious.

For systematic reasons, it is worth remembering that the concept of the unconscious covers a broad spectrum of meanings. For example, neuroscience identifies the unconscious mainly in a descriptive manner, as that which is not conscious, or as the region beneath the threshold of perception, or as associative or autobiographical memory. And in object relations psychoanalysis, the unconscious tends to be regarded in a dynamic way, as a concept for what the ego has repressed. Still, the unconscious has another meaning as part of the concept of topography: the unconscious as a system, the *Ucs*, which Freud in *A Note on the Unconscious in Psycho-Analysis/ Einige Bemerkungen über den Begriff des Unbewußten in der Psychoanalyse* (1912) called the third and most important meaning of the unconscious (Stud. III, 36). Today this latter meaning appears to have largely been forgotten, even though it is of fundamental importance for the concept of the psyche. This topographical meaning gives the unconscious an aspect that differentiates it clearly from the neuroscientific definition—a touch of radical elusiveness, alterity, dissociation, mystery, and inscrutability, mostly silent and directed toward the self. Here the unconscious is perceived by the individual as a foreign object, something alien. Moreover, providing for continued existence and taking precautions to guarantee survival are not the primary interests of the unconscious from this perspective. To complicate matters further, the processes in this dimension also occur beyond conscious mediation.

It is easy to imagine that this dimension makes it quite difficult to classify psychoanalysis scientifically. All the same, Freud was probably onto this dimension from the beginning, at least by the time he wrote his 1891 study on aphasia. For example, in the *Interpretation of Dreams/Die Traumdeutung* (1900) he calls the unconscious the "other scene" (S.E. V, 535), a fictional place. By using the impersonal pronoun for the 'Id' of the second topography (in German *Es*), he also emphasizes its characteristic of autonomy, which is very important. He continues to declare the Id to be the darkest, oldest part of the soul. It lends its atmosphere to all other unconscious pieces of evidence. The unconscious thus becomes the Black Continent, our internal Africa, and Freud adds that it determines our lives in their totality. Freud remained strongly convinced of this notion of the unconscious, so much so that by 1938, in *An Outline of Psycho-Analysis,* he repeated that a part of the unconscious would always remain inaccessible. In an almost Lacanian manner, he also claims here that the real always remains unrecognizable. Freud thus founded the concept of the psyche, in the psychoanalytical sense, on inaccessibility and foreignness.

The idea of the abyss, which I have touched upon only briefly here, has an enormous impact. It influences all of psychoanalysis' theoretical concepts from repression and drives to the body and narcissism, and not least to the subject itself—a concept that still exists in psychoanalysis but not in neuroscience. Nonetheless, the question arises as to what this odd positing of the placeless unconscious is supposed to mean? Does it make Freud a romantic, an opponent of the Enlightenment, or even a mystic? Is this positing itself epistemically deficient and misguided? Hasn't all this been brought up to date by neuroscience? I think not. On the contrary, I want to argue that Freud's hypothesis of the placeless psyche is part of a programmatic system, a paradoxical program. In the context of this program, Freud proves to be the author of a paradoxically constructed psyche whose individual parts do not add up or even fit together.[1] In doing so, Freud propounds the concept of a psyche, of a psychical apparatus, whose tendencies to block itself are stronger than its tendencies of constructive self-organization. And by no means does it self-organize. Yet Freud had already determined all of this from the beginning.

And this is the crux of the matter: the leap across the crack in the psyche that was already hinted at in his *Project for a Scientific Psychology/Entwurf einer Psychologie* (1895) in the form of the psychological apparatus's principle of inertia and the internal movements of the energy processes counteracting each other. That is the solution to the puzzle that Freud, to his displeasure, was unable to solve as long as he followed the hypotheses of "psycho-physical parallelism." But when he finally elaborated the concept of the *unconscious* in 1915, he found the missing link that was able to bridge the fragile logic of consciousness. It explains the effects of hypnotic symptoms and defines the relationship to the anatomical-biological body. From that point on, the body becomes characterized by the gap in the *unconscious* and is seen as rooted in the concept of the *drive*.[2] The concept of the unity of body

---

[1] Psychoanalysis does not ask about the causes but about the effects. It asks about the causality of effects, not of causes.

[2] Psychologically speaking, the *idea* of the body is what matters, not the anatomical body. The idea of the body follows the laws of language—it is the spoken body.

and soul split open, and the closed system was cracked by a *mysterious leap* across the chasm. Biology was subsequently placed in the shadow of the biological facts.

In order to conceptualize Freudian psychoanalysis, we can state the following: The psyche is an ambivalent entity. On the one hand, it is not capable of consciousness or generalizations. From this angle, it exists only as the existential-logical necessity of a hypothesis. On the other hand, it builds upon those aspects of the unconscious— the dynamic and descriptive unconscious—that are easily capable of consciousness and generalization. Freud accords meaning to both sides: the irreducibly topographical side as well as the manifestable side. On top of that, he maintains that they are interlinked. In other words, the hypothetical lack of foundation of the psychical, this *placeless place* (the Id), affects psychical manifestations and lessens their validity, while the general order is responsible for bridging the gap to the unattainable Id.

I should say a word or two about the status of neuroscience within the history of science. During his time as a neurologist, Freud had already recognised that there was a connection between the unconscious and what he called linguistic trajectory of discharge, that is, language, a connection from which he derived the method of the "talking cure." Following and accentuating Freud's discovery Lacan later developed the theorem of a symbolic order of the psyche which is basically construed according to (formal) linguistic principles. In this context, we should not overlook the fact that defining fundamental principles comparatively in terms of formal language is not limited to psychoanalytic theorems and phenomena alone. Indirectly such a definition also affects neuroscientific discoveries and assertions. Neuroscientific methods like image producing procedures heavily rely on digital methods of processing images like MRIs which examine, measure and represent nerve tissue. However, the analysis of the date such procedures yield is dependant on algorithms that in turn rely on the numeric symbolic language of mathematics. Even neuroscience in using these techniques and thus relying on numerical rather than alphabetical language in the last resort is dependent on language. If we furthermore take into account that these "technical images" are translated into "manmade" linguistic hypotheses we cannot avoid the conclusion that neuroscience does not simply produce verifiable and objective propositions about the functioning neuronal chains but should rather restrict itself to statements about relative probabilities and that it is crucially influenced by the symbolic language of mathematics.

For this reason, the conclusion that neuroscience ought to be thought of as providing evidence for the embodiment of a (regularly functioning) spirit or for the objective basis of the psyche (E. Kandel, G. Edelman, A. R. Damasio) must be considered an epistemic self-misunderstanding. I would therefore like to modify my initial assertion that the levels of the laws in neuroscience and psychoanalysis are comparable. To the contrary, I would like to claim now that the laws postulated by neuroscientists about the objectivity of "gene expressions" or synaptic connections are not of the same kind of laws as in psychoanalysis. With the history of psychoanalysis in mind—in particular Freud's distancing himself from the greatest neurologist of the nineteenth century, Jean-Martin Charcot—I am convinced that there is an epistemic leap required in order to cross the gap between the empirical objectivity of neurons and the symbolic order of the psyche.[3] Faced with Charcot's scientification of hysteria, it dawned on Freud that one dimension had been obsessively omitted: the dimension of the particular, which is to say, of the individual subject and its unconscious makeup. It becomes apparent just how fragile the notion of singularity is when one recognizes that, today just as then, some people champion the opinion that psychoanalytical research on individual cases, as productive as it may be, cannot be imagined without support from independent and objective methods.

Meanwhile, one might wonder whether Freud's discovery of the singularity of the unconscious subject still has its place in today's neurological theories, regardless of whether the theories deal with identity or are conceived of as reductive and materialistic. Perhaps the subject with its desires and dreams has indeed played its part and is now present only as a living construction in mice, fruit flies, and snails. Nevertheless, the suspicion arises that this relegation of the unconscious would rob the human individual of a part of his or her liveliness and vitality. For Freud that was precisely what made a person happiest and also what was the most daring and dangerous aspect of existence, for the unconscious had no particular purpose. What are the consequences of robbing a person of that part which psychoanalysis had reserved as the place for the dead within a part of the living?

I would like to conclude with the quotation I used for the title of my article. It is, of course, from Freud's *Introductory Lectures on Psycho-Analysis* (1917):

> The medulla oblongata is a very serious and lovely object. I remember quite clearly how much time and trouble I devoted to its study, many years ago. To-day, however, I must remark that I know nothing that could be of less interest to me for the psychological understanding of anxiety than a knowledge of the path of the nerves along which its excitations pass. (S.E. XV, 393)

(English translation by Sandra Lustig)

References

Damasio, A. R. (1994). *Descartes' error. Emotion, reason, and the human brain.* New York: Quill.
Du Bois-Reymond, E. (1887). Über die Lebenskraft. In E. Du Bois-Reymond (Ed.), *Reden von Emil du Bois-Reymond, BD II* (pp. 1–28). Leipzig: Veit & Co. (Original work published 1848)
Freud, S. (1895). *Project for a scientific psychology.* S.E. I.
Freud, S. (1900). *Interpretation of dreams.* S.E. V.
Freud, S. (1912). Einige Bemerkungen über den Begriff des Unbewußten in der Psychoanalyse. Stud. III.
Freud, S. (1917). *Introductory lectures on psychoanalysis/Vorlesungen zur Einführung in die Psychoanalyse.* S.E. XV.
Freud, S. (1940). *An outline of psycho-analysis.* S.E. XXIII.
Grünbaum, A. (1988). *Die Grundlagen der Psychoanalyse: Eine philosophische Kritik.* Stuttgart: Reclam.

---

[3] As Wolfgang Leuschner writes, Freud did finally turn the doctor's armchair away from the patient by 180° (Leuschner 1992).

Habermas, J. (1977). *Erkenntnis und Interesse*. Frankfurt: Suhrkamp Verlag.
Lacan, J. (2006). Subversion of the subject and the dialectics of desire in the Freudian unconscious. In J. Lacan, *Écrits*. (B. Fink, Trans.) New York: Norton.
Leuschner, W. (1992). Introduction. In S. Freud (Ed.), *Zur Auffassung der Aphasien* (pp. 7–35). Frankfurt: Fischer Verlag.
Libet, B. (2004). *Mind time: The temporal factor in consciousness*. Cambridge, MA: Harvard Press.
Pauen, M. (2001). *Grundprobleme der Philosophie des Geistes*. Frankfurt: Fischer Verlag.
Ricoeur, P. (1970). *Freud & philosophy: An essay on interpretation*. New Haven: Yale University Press.
Seifert, E. (2008). *Seele- Subjekt- Körper*. Gießen: Psychosozial Verlag.
Singer, W. (1990). The formation of cooperative cell assemblies in the visual cortex. *Journal of Experimental Biology, 153*, 177–197.

On the Authors

François Ansermet is currently Chair of the Department of Psychiatry, Professor of Child and Adolescent Psychiatry at the University of Geneva. His research line is on developmental stress and psychopathology, specifically in the field of perinatal stress and prematurity, and on psychological and ethical implications of new advances in perinatal biotechnology (esp. assisted reproductive therapy, prenatal predictive medicine and gender attributions in cases of intersexuality). His interdisciplinary research in the field of neurosciences and psychoanalysis focuses on the role of neuronal plasticity in the psychoanalytic process. He created with Professor Pierre Magistretti the Agalma Foundation (www.agalma.ch) for the study of the unconscious through the lens of contemporary neurosciences. His works include: *Biology of Freedom: Neural Plasticity, Experience, and the Unconscious* (co-author, 2007), *Neurosciences et psychanalyse: une rencontre autour de la singularité* (co-author, 2010), *Les énigmes du plaisir* (co-author, 2010), *A chacun son cerveau. Plasticité neuronale et inconscient* (co-author, 2011), *Clinique de l'origine* (2012), *Autisme: à chacun son génome* (co-author, 2013), and Memory reconsolidation, trace reassociation and the Freudian unconscious. In: C.M. Alberini (Ed.), *Memory Reconsolidation*. Amsterdam: Elsevier/Academic Press (2013, pp. 293–312, co-author).

Tamara Fischmann is a psychoanalyst (German Psychoanalytic Association/International Psychoanalytic Association; DPV/IPA) and scientific researcher at the Sigmund-Freud-Institut, Frankfurt am Main. She specializes in experimental dream research, in neuropsychoanalysis and in methodology in psychoanalytic research. Selected publications: Trauma, dream and psychic change in psychoanalyses: a dialogue between psychoanalysis and the neurosciences. In: *Front. Hum. Neurosci* (7/2013, pp. 877); Changes in Dreams of chronic depressed patients. The Frankfurt fMRI/EEG Depression Study (FRED). In: P. Fonagy, H. Kächele, M. Leuzinger-Bohleber, D. Tyalor (Eds.), *The Significance of Dreams—Bridging Clinical and Extraclinical Research in Psychoanalysis* (co-author, 2012, pp. 159–183); *Ethical Dilemmas in Prenatal Diagnosis* (co-editor, 2011); Aufsuchende Psychoanalyse in der Frühprävention. Klinische und extraklinisch-empirische Studien. In: *Frühe Bildung* (co-author, 2/2013, pp. 72–83); Early prevention in day-care centres with children at risk—the EVA research project. In: Robert N. Emde, Marianne Leuzinger-Bohleber (Eds.), *Early Parenting and Prevention of Disorder: Psychoanalytic Research at Interdisciplinary Frontiers* (co-author, 2014, pp. 242–259).

Ulrike Kadi is Assistant Professor in the Department of Psychoanalysis and Psychotherapy at the Medical University of Vienna and University Lecturer in the Department of Philosophy at the University of Vienna, Austria. Her research interests include poststructuralism, phenomenology and Lacanian psychoanalysis. Selected publications: *Sinn macht Unbewusstes, Unbewusstes macht Sinn* (co-author, 2005); *Wahn: Philosophische, psychoanalytische und kulturwissenschaftliche Perspektiven* (co-editor, 2012); Unmögliche Trennung. In: Stephan Doering, Heidi Möller (Eds.), *Mon Amour trifft Pretty Woman. Liebespaare im Spielfilm* (2014, pp. 143–157).

Marianne Leuzinger-Bohleber is President of the Sigmund-Freud-Institut, Frankfurt am Main, Professor for Psychoanalysis at the University of Kassel, President of the research and academic committee of the German Psychoanalytic Association (DPV), Vice Chair for Europe, International Research Boards of the International Psychoanalytical Association, Visiting Professor at the University College London, and a member of the Action Group Neuro-Psychoanalysis. She focuses on clinical and empirical research of psychoanalysis, on psychoanalytic developmental psychology, and interdisciplinary research between psychoanalysis and related fields (esp. cognitive science, neurosciences, literary studies). Recent works include *Embodiment. Ein innovatives Konzept für Entwicklungsforschung und Psychoanalyse* (co-editor, 2013); *Chronische Depression. Verstehen—Behandeln—Erforschen* (co-editor, 2013); *Psychoanalyse—die Lehre vom Unbewussten. Geschichte, Klinik und Praxis* (co-editor, 2014).

Pierre J. Magistretti is Professor of Neuroscience at the Brain Mind Institute at the École Polytechnique Fédérale de Lausanne (EPFL), Dean of the Division of Biological and Environmental Sciences and Engineering at King Abdullah University of Science and Technology, Thuwal, Saudi Arabia (KAUST), and President of the International Brain Research Organization (IBRA). His interdisciplinary research of neurosciences and psychoanalysis focuses on cellular and molecular mechanisms of the coupling between neuronal activity and energy consumption by the brain. Recent publications: *Les énigmes du plaisir* (co-author, 2011); *A chacun son cerveau. Plasticité neuronale et inconscient* (co-author, 2011); *Gli enigmi del piacere* (co-author, 2012).

Gerhard Scharbert lectures at the Institut für Kulturwissenschaft at the Humboldt-Universität zu Berlin. From 2008 to 2011 he worked as a researcher in the project

"Freud und die Naturwissenschaften um 1900 und um 2000" at the Berlin Center for Literary and Cultural Research (Zentrum für Literatur- und Kulturforschung Berlin). His interests are the historical and current interdependencies between neurosciences, psychiatry and aesthetics, psychopathology and linguistic theory, and the ontological foundation of poetics. His works include *Das Locked-in-Syndrom: Geschichte, Erscheinungsbild, Diagnose und Chancen der Rehabilitation* (2010), *Dichterwahn. Über die Pathologisierung von Modernität* (2010), *Freuds Referenzen* (co-author, 2011).

Edith Seifert is a practicing psychoanalyst and lectures at the Leopold Franzens University Innsbruck. She is a founding member of the Psychoanalytic Salon, Berlin. Her research interests include differences and common grounds of neurosciences, neuropsychology and psychoanalysis, Lacanian psychoanalysis and female sexuality. Selected publications: *Was will das Weib? Zu Begehren und Lust bei Freud und Lacan* (1987); *Perversion der Philosophie. Lacan und das unmögliche Erbe des Vaters* (editor, 1992); *Seele—Subjekt—Körper: Freud mit Lacan in Zeiten der Neurowissenschaft* (2008); *Die Wette auf das Unbewußte oder Was Sie schon immer über Psychoanalyse wissen wollten* (co-author, 2008).

Mark Solms is Professor in Neuropsychology at the University of Cape Town and Groote Schuur Hospital, and President of the South African Psychoanalytical Association. He is a member of the British Psychoanalytical Society, and was awarded Honorary Membership of the New York Psychoanalytic Society in 1998. His research focuses on brain mechanisms of dreaming, and psychoanalytic methods and theories in contemporary neuroscience. He has published more than 250 chapters and articles in both neuroscientific and psychoanalytic journals, and is the editor of the Revised Standard Edition of the Complete Psychological Works of Sigmund Freud (24 vols.) and the forthcoming Complete Neuroscientific Works of Sigmund Freud (4 vols.). Selected book publications: *The Neuropsychology of Dreams: A Clinical-Anatomical Study* (1997), *Clinical Studies in Neuropsychoanalysis* (2000) and *The Brain and the Inner World: An introduction to the Neuroscience of Subjective Experience* (co-author, 2002).

Oliver H. Turnbull is Pro Vice Chancellor (Teaching and Learning) at Bangor University, Wales, and Professor of Psychology and Neuropsychology. He is founder member and secretary of the International Neuropsychoanalysis Society, a member of the British Neuropsychological Society, Fellow of the Royal Society of Medicine and of the British Psychological Society. His research interests include the link between psychoanalysis and neuropsychology, trauma and brain injuries, brain physiology and emotions. Selected publications: *The Brain and the Inner World. An Introduction to the Neuroscience of Subjective Experience* (co-author, 2002); Reports of intimate touch: Erogenous zones and somatosensory cortical organization. In: *Cortex* (co-author, 53/2013, pp. 146–154); Cognitive vulnerability to bipolar disorder in offspring of parents with bipolar disorder. In: *British Journal of Clinical Psychology* (co-author, online publication, Apr. 2014).

Sigrid Weigel is Director of the Center for Literary and Cultural Research (Zentrum für Literatur- und Kulturforschung) in Berlin. She has held professorships at the universities of Hamburg, Zurich, and at the Technische Universität Berlin, and was visiting professor in Basel, Berkeley, Cincinatti, Harvard, Princeton, and Stanford. Her research focuses on the way cultural sciences position themselves between the humanities and the natural sciences. She is currently working on a history of compassion and on image theories. Book publications include: *Literatur als Voraussetzung der Kulturgeschichte. Schauplätze von Shakespeare bis Benjamin* (2004); *Genea-Logik. Generation, Tradition und Evolution zwischen Kultur- und Naturwissenschaften* (2006); *Heine und Freud. Die Enden der Literatur und die Anfänge der Kulturwissenschaft* (ed., 2010); *Walter Benjamin—Images, the Creaturely, and the Holy* (2013); *Gesichter* (ed., 2013); *Grammatologie der Bilder* (2015), *Empathy. A Neurobiological Capacity and its Cultural and Conceptual History* (co-author, 2015).

Yoram Yovell is Associate Professor for Neuroscience, Psychiatry and Psychoanalysis; Co-Director of the Institute for the Study of Affective Neuroscience (ISAN) at the University of Haifa, Israel; training and supervising psychoanalyst at the Israel Psychoanalytic Institute, Jerusalem; and founding member of the International Neuropsychoanalysis Society. His research focuses on pharmacological treatment of suicidality, and neurobiology of memory disturbances following psychological trauma. His writings include: *Der Feind in meinem Zimmer. Geschichten aus der Psychotherapie* (2004); *Liebe und andere Krankheiten* (2006); Characteristics of attachment style in woman with dyspareunia (co-author, 2011). In: *J Sex Marital Ther.* 2011; 37(1): pp. 1–16; Preclinical modeling of primal emotional affects (SEEKING, PANIC and PLAY): Gateways to the development of new treatments for depression" (co-author, 2014) in: *Psychopathology*. 22.10.14: DOI: 10.1159/000366208.

The manufacturer's authorised representative in the EU is Springer Nature Customer Service Centre GmbH, Europaplatz 3, 69115 Heidelberg, Germany. If you have any concerns regarding our products, please contact ProductSafety@springernature.com

Printed and bound by CPI Group (UK) Ltd, Croydon, CR0 4YY

23/03/2026

02076379-0004